BIBLIOGRAPHIC RECORD TARGET

Graduate Library
University of Michigan

Preservation Office

Storage Number: ______________

ABN2643
UL FMT B RT a BL m T/C DT 07/18/88 R/DT 07/18/88 CC STAT mm E/L 1
010: : |a 07038569
035/1: : |a (RLIN)MIUG86-B42066
035/2: : |a (CaOTULAS)160031819
040: : |c CtY |d CtY |d MiU
050/1:0 : |a QA501 |b .W59
100:1 : |a Whitehead, Alfred North, |d 1861-1947.
245:04: |a The axioms of descriptive geometry, |c by A.N. Whitehead.
260: : |a Cambridge, |b University press, |c 1907.
300/1: : |a viii, 74 p. |b diagrs. |c 22 cm.
440/1: 0: |a Cambridge tracts in mathematics and mathematical physics. |v no. 5
650/1: 0: |a Geometry, Descriptive
998: : |c RSH |s 9124

Scanned by Imagenes Digitales
Nogales, AZ

On behalf of
Preservation Division
The University of Michigan Libraries

Date work Began: ______________
Camera Operator: ______________

Cambridge Tracts in Mathematics
and Mathematical Physics

GENERAL EDITORS
J. G. LEATHEM, M.A.
E. T. WHITTAKER, M.A., F.R.S.

No. 5

THE AXIOMS OF DESCRIPTIVE GEOMETRY

by
A. N. WHITEHEAD, Sc.D., F.R.S.

Cambridge University Press Warehouse
C. F. CLAY, Manager
London: Fetter Lane, E.C.
Glasgow: 50, Wellington Street
1907

Price 2s. 6d. net

Cambridge Tracts in Mathematics and Mathematical Physics

No. 5

The Axioms of Descriptive Geometry.

CAMBRIDGE UNIVERSITY PRESS WAREHOUSE,
C. F. CLAY, MANAGER.
London: FETTER LANE, E.C.
Glasgow: 50, WELLINGTON STREET.

Leipzig: F. A. BROCKHAUS.
New York: G. P. PUTNAM'S SONS.
Bombay and Calcutta: MACMILLAN AND CO., LTD.

Cambridge Tracts in Mathematics and Mathematical Physics

No. 5

The Axioms of Descriptive Geometry.

CAMBRIDGE UNIVERSITY PRESS WAREHOUSE,
C. F. CLAY, MANAGER.
London: FETTER LANE, E.C.
Glasgow: 50, WELLINGTON STREET.

Leipzig: F. A. BROCKHAUS.
New York: G. P. PUTNAM'S SONS.
Bombay and Calcutta: MACMILLAN AND CO., LTD.

THE AXIOMS OF DESCRIPTIVE GEOMETRY

by

A. N. WHITEHEAD, Sc.D., F.R.S.

Fellow of Trinity College.

CAMBRIDGE:

at the University Press

1907

Cambridge:
PRINTED BY JOHN CLAY, M.A.
AT THE UNIVERSITY PRESS.

PREFACE.

THIS tract is written in connection with the previous tract, No. 4 of this series, on Projective Geometry, and with the same general aims. In that tract, after the statement of the axioms, the ideas considered were those concerning harmonic ranges, projectivity, order, the introduction of coordinates, and cross-ratio. In the present tract, after the statement of the axioms, the ideas considered are those concerning the association of Projective and Descriptive Geometry by means of ideal points, point to point correspondence, congruence, distance, and metrical geometry. It has been my object in both tracts to extend the investigations just far enough to assure the reader that the whole of Geometry is really secured by the axioms stated. My hopes for a comparative freedom from typographical errors are based upon my experience of the excellence of the University Press.

A. N. W.

CAMBRIDGE.
March, 1907.

CONTENTS.

CHAPTER I.

FORMULATIONS OF THE AXIOMS.

1. THE general considerations which must govern a mathematical nvestigation on the foundations of Geometry have been explained in hapter I of the previous tract of this series, on the Axioms of rojective Geometry*. It is explained there that 'Descriptive eometry' is here used as a generic term for any Geometry in which wo straight lines in a plane do not necessarily intersect. Also it is ointed out that the purely classificatory portions of a Descriptive eometry are clumsy and uninteresting, and that accordingly the idea f order is introduced from the very beginning.

There are three main ways by which this introduction of order an be conveniently managed. In one way, which is represented by eano's axioms given below (§§ 3—6), the *class* of points which lie *etween* any two points is taken as a fundamental idea. It is then sy to define the class of points collinear with the two points and ing *beyond* one of them. Thus these three classes of points, namely e two classes lying beyond the two points respectively and the class ing between the two points, together with the two points themselves rm the straight line defined by the two points. Then a set of axioms the straight line are required, concerned with the idea of 'between,' d also axioms are required respecting coplanar lines.

Another way, which was pointed out by Vailati† and Russell‡, is conceive a straight line as essentially a serial relation involving two rms. The whole field of such a relation, namely the terms which are us ranged in order by it, forms the class of points on the straight e. Thus the Geometry starts with the fundamental conception of a

* In the sequel this tract will be referred to as 'Proj. Geom.'

† Cf. *Rivista di Matematica*, vol. IV.

‡ Cf. *Principles of Mathematics*, § 376.

class of *relations*. The axioms of the straight line are the axioms which secure that each of these relations is a serial relation. The points are the entities occurring in the fields of any of these relations. The axioms of the plane are the same as in the previous mode of development.

The third way, recently developed by Prof. O. Veblen*, is to consider the science of Descriptive Geometry as the study of the properties of one single three-termed relation of order. The entities forming the field of this relation are the points. When this relation holds between three points A, B, and C, it is said that 'the points A, B, and C, are in the linear order ABC.' This method of conceiving the subject results in a notable simplification, and combines advantages from the two previous methods. Veblen's axioms will be stated in full (cf. § 8).

2. The enunciation of the axioms of Descriptive Geometry, which is given in the sections (§§ 3—6) immediately following, is that due to Peano †. His formulation is based upon that of Pasch ‡, to whom is due the first satisfactory systematic exposition of the subject. The undefined fundamental ideas are two in number, namely that of a class of entities called 'points,' and that of the 'class of points lying between any two given points.' It has already been explained§ that this undetermined class of points is in fact any class of entities with inter-relations, such that the axioms are satisfied when considered as referring to them.

The symbol AB will represent the class of points lying between the points A and B. This class will be called the segment AB.

The first group of axioms, eleven in number, secure the ordinary properties of a straight line with respect to the order of points on it, and also with respect to the division of a line|| into three parts by any

* Cf. *A System of Axioms for Geometry*, Trans. of the Amer. Math. Soc., vol. v., 1904.

† Cf. *I principii di Geometria*, Turin, 1889. These axioms are repeated by him in an article, *Sui fondamenti della Geometria*, Rivista di Matematica, vol. IV. 1894. In this latter article the minute mathematical deductions are omitted, and their place is taken by valuable observations on the main points to be considered. Also a treatment of congruence is given which does not appear in the earlier tract. This article should be studied carefully by every student of the subject.

‡ Cf. *Vorlesungen über neuere Geometrie*, Leipzig, 1882. This treatise is the classic work on the subject.

§ Cf. Proj. Geom. § 2.

|| Note that 'line' will be habitually used for 'straight line.'

two points on it, and into two parts by any single point on it. The Dedekind property* is not secured by them, but, compactness† is secured by axiom IV.

3. Peano's axioms of the straight line are as follows:

I. There is at least one point.

II. If A is any point, there is a point distinct from A.

III. If A is a point, there is no point lying between A and A.

It follows that the class AA possesses no members.

IV. If A and B are distinct points, there is at least one point lying between A and B.

Thus the class AB is not the null class.

V. If the point C lies between A and B, it also lies between B and A.

It easily follows that the classes AB and BA are identical.

VI. The point A does not lie between the points A and B.

Thus the class, or segment, AB does not include its end-points A and B.

Definition. If A and B are points, the symbol $A'B$ represents the class of points, such as C, with the property that B lies between A and C. Thus $A'B$ is the prolongation of the line beyond B, and $B'A$ is its prolongation beyond A.

VII. If A and B are distinct points, there exists at least one member of $A'B$.

VIII. If A and D are distinct points, and C is a member of AD, and B of AC, then B is a member of AD.

IX. If A and D are distinct points, and B and C are members of AD, then either B is a member of AC, or B is identical with C, or B is a member of CD.

X. If A and B are distinct points, and C and D are members of $A'B$, then either C is identical with D, or C is a member of BD, or D is a member of BC.

XI. If A, B, C, D are points, and B is a member of AC, and C of BD, then C is a member of AD.

Definition. The straight line possessing A and B, symbolized by str (A, B), is composed of the three classes $A'B$, AB, $B'A$, together with the points A and B themselves.

Then by the aid of the previous axioms the usual theorems,

* Cf. Proj. Geom. § 19, and § 9 of the present tract.

† Cf. Proj. Geom. § 16.

excluding the Dedekind property, respecting the order of points on a line can be proved. Also any two points are both contained by one and only one line.

4. Peano uses the following useful notation which is an extension of his notation for segments and prolongations. If A is a point and u is a class of points, then Au is the class of points lying on the segments between A and points of u, and $A'u$ is the class of points on the prolongations of these segments beyond the points of u.

Then in conformity with this notation the seven regions into which a plane is divided by three lines are as in the figure.

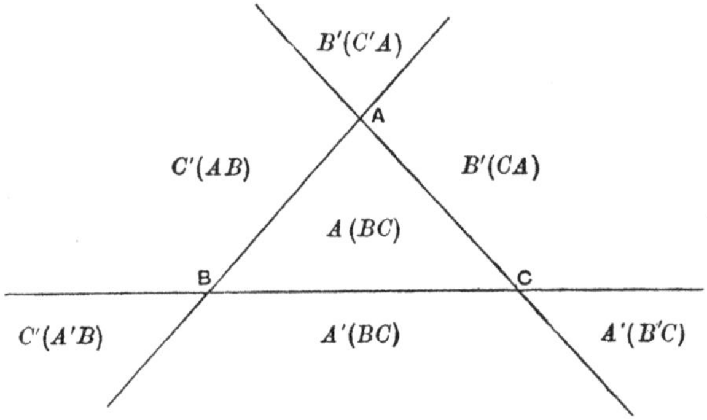

The plane determined by the three non-collinear points A, B, C—written ple (A, B, C)—is defined to be the class of points consisting of the points on the three lines str (BC), str (CA), and str (AB), and of the points in the seven regions $A\,(BC)$, $A'\,(BC)$, $B'\,(CA)$, $C'\,(AB)$, $A'\,(B'C)$, $B'\,(C'A)$, $C'\,(A'B)$.

5. Three axioms are required to establish the Geometry of a plane.

XII. If r is a straight line, there exists a point which does not lie on r.

Note that it would be sufficient to enunciate this axiom for one straight line.

XIII. If A, B, C are three non-collinear points, and D lies on the segment BC, and E on the segment AD, there exists a point F on both the segment AC and the prolongation $B'E$ (cf. fig. i, p. 5).

XIV. If A, B, C are three non-collinear points, and D lies on the segment *BC*, and *F* on the segment *AC*, there exists a point *E* lying on both the segments *AD* and *BF* (cf. fig. ii).

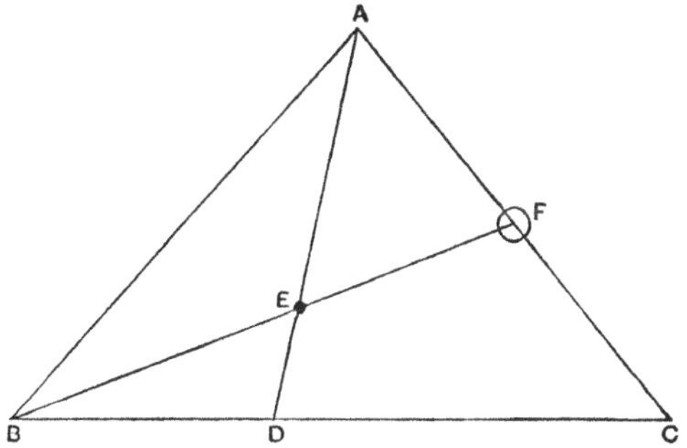

Fig. i.

With these axioms all the usual properties of the division of a plane by a line, and of the inside and outside of a plane closed figure,

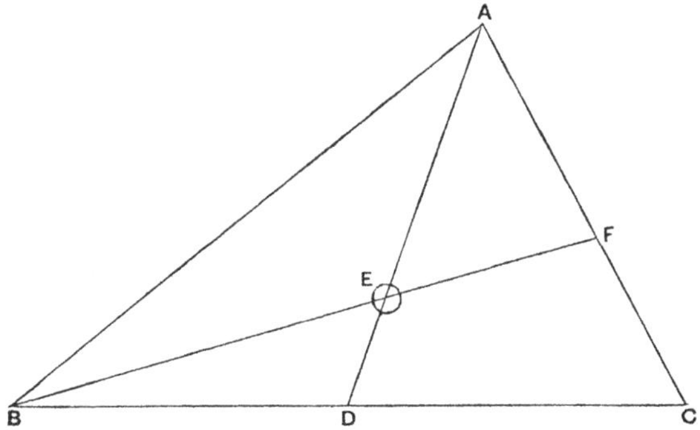

Fig. ii.

can be proved. Thus if *ABC* form a triangle and a coplanar line intersect the segment *BC*, it must intersect one and only one of the segments *CA* and *AB*.

Also any three non-collinear points lie in one and only one plane; and the line determined by any two points lying in a plane lies entirely in that plane. But, as the case of Euclidean Geometry shews, we cannot prove from these axioms that any two lines in a plane intersect.

6. For three-dimensional Geometry two other axioms are required.

XV. A point can be found external to any plane. The enunciation of this axiom can be restricted to a particular plane.

XVI. Given any plane p, and any point A outside it, and any point Q on it, and any point B on the prolongation $A'Q$, then, if X is any other point, either X lies on the plane p, or AX intersects the plane p, or BX intersects the plane p.

The annexed figure illustrates the axiom, the points X_1, X_2, X_3 being positions of X which illustrate the three alternatives contemplated in the axiom. Thus X_1 lies on the plane p; X_2 lies on the

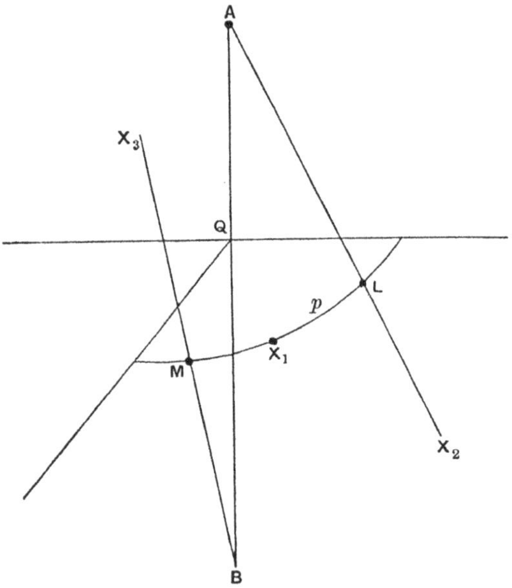

same side of p as B, so that AX_2 must cut p in some point L; X_3 lies on the same side of p as A, so that BX_3 must cut p in some point M.

Axiom XVI secures the limitation to three dimensions, and the division of space by a plane. It can also be proved from the axioms that, if two planes intersect in at least one point, they intersect in a straight line.

7. A point will be said to divide a line into two half-rays which emanate from it.

A line will be said to divide a plane into two half-planes which are bounded by it.

A plane will be said to divide space into two half-spaces which are bounded by it.

A sheaf of lines is a complete set of coplanar lines concurrent at one point (the vertex). A sheaf of half-rays is a complete set of coplanar half-rays emanating from one point (the vertex).

A bundle of lines is a complete set of lines concurrent at one point (the vertex). A bundle of half-rays is a complete set of half-rays emanating from one point (the vertex).

If p, q, r are three half-rays belonging to a sheaf of half-rays, then r is said to 'lie between' p and q, if points A and B can be found on p and q respectively, such that the segment AB intersects r.

It can be proved that if r lies between p and q, then p does not lie between r and q.

The complete set of planes through a given line (the axis) is called a sheaf of planes. The axis divides each plane into two half-planes. These half-planes form a sheaf of half-planes.

If p, q, r are three half-planes belonging to a sheaf of half-planes, then r is said to 'lie between' p and q, if points A and B can be found on p and q respectively, such that the segment AB intersects r.

It can be proved that if r lies between p and q, then p does not lie between r and q.

The theorems indicated in this and in the preceding sections, and allied theorems, are not always very easy to prove. But their proofs depend so largely upon the particular mode of formulation of the axioms, that it would be outside the scope of this tract to enter into a consideration of them. In the sequel we shall assume that the whole class of theorems of the types, which have been thus generally indicated, can be proved from the axioms stated.

8. Formulations of the axioms of Descriptive Geometry have also been given by Hilbert*, and by E. H. Moore†, and by B. Russell‡, and by O. Veblen§. Veblen's memoir represents the final outcome of these successive labours, and his formulation will be given now. The axioms are stated in terms of 'points' and of a relation among three points called 'order.' Points and order are not defined.

I. There exist at least two distinct points.

* *Grundlagen der Geometrie*, Leipzig, 1899, English Translation by E. J. Townsend, Chicago, 1902.

† *On the Projective Axioms of Geometry*, Trans. of the Amer. Math. Soc., vol. III., 1902.

‡ *The Principles of Mathematics*, Cambridge, 1903, ch. XLVI.

§ *A System of Axioms for Geometry*, Trans. of the Amer. Math. Soc., vol. V., 1904.

II. If the points A, B, C are in the order ABC, they are in the order CBA.

III. If the points A, B, C are in the order ABC, they are not in the order BCA.

IV. If the points A, B, C are in the order ABC, then A is distinct from C.

V. If A and B are any two distinct points, there exists a point C such that A, B, C are in the order ABC.

Definition 1. The line AB $(A \neq B)$ consists of A and B, and of all points X in one of the possible orders ABX, AXB, XAB. The points X in the order AXB constitute the 'segment' AB. A and B are the 'end-points' of the segment, but are not included in it.

VI. If points C and D $(C \neq D)$ lie on the line AB, then A lies on the line CD.

VII. If there exist three distinct points, there exist three points A, B, C not in any of the orders ABC, BCA, or CAB.

Definition 2. Three distinct points not lying on the same line are the 'vertices' of a 'triangle' ABC, whose sides are the segments AB, BC, CA, and whose 'boundary' consists of its vertices and the points of its sides.

VIII. If three distinct points A, B, C do not lie on the same line, and D and E are two points in the orders BCD and CEA, then a

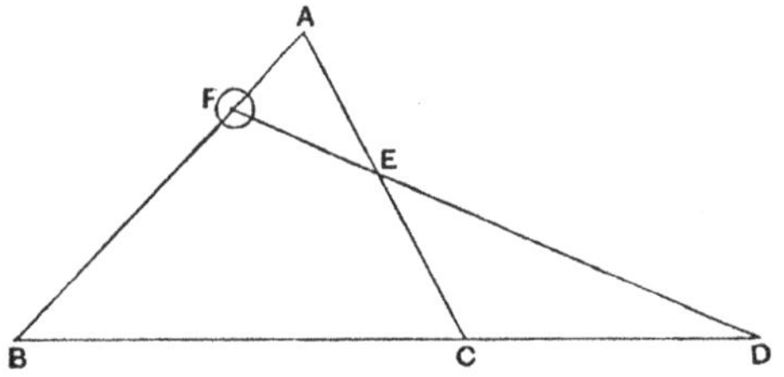

point F exists in the order AFB and such that D, E, F lie on the same line.

Definition 5. A point O is 'in the interior of' a triangle, if it lies on a segment, the end-points of which are points of different sides of a triangle. The set of such points O is 'the interior' of the triangle.

Definition 6. If A, B, C form a triangle, the 'plane' ABC consists of all points collinear with any two points of the sides of the triangle.

IX. If there exist three points not lying in the same line, there exists a plane ABC such that there is a point D not lying in the plane ABC.

Definition 7. If A, B, C, and D are four points not lying in the same plane, they form a 'tetrahedron' $ABCD$, whose 'faces' are the interiors of the triangles ABC, BCD, CDA, DAB, whose 'vertices' are the four points A, B, C, and D, and whose 'edges' are the segments AB, BC, CD, DA, AC, BD. The points of faces, edges, and vertices constitute the 'surface' of the tetrahedron.

Definition 8. If A, B, C, D are the vertices of a tetrahedron, the space $ABCD$ consists of all points collinear with any two points of the faces of the tetrahedron.

X. If there exist four points, neither lying in the same line, nor lying in the same plane, there exists a space $ABCD$, such that there is no point E not collinear with two points of the space $ABCD$.

The above axioms of Veblen are equivalent to the axioms of Peano which have been previously given. Both Peano and Veblen give an axiom securing the Dedekind property (cf. § 9). Also Veblen gives an axiom securing the 'Euclidean' property (cf. § 10).

9. Dedekind's original formulation* of his famous property applies directly to the case of a descriptive line and is as follows:

"If all points of the straight line fall into two classes such that every point of the first class lies to the left of every point of the second class, then there exists one and only one point which produces this division of all points into two classes, this severing of the straight line into two portions."

It is of course to be understood that the dividing point itself belongs to one of the two classes.

It follows immediately that the boundary of a triangle consists of points in a compact closed order possessing the Dedekind property as already formulated for closed series†.

This definition may be repeated here to exhibit its essential independence of the special definition of projective segments upon which the previous formulation rests.

Let A, B, C be any three points of a closed series. Then by

* Cf. his *Continuity and Irrational Numbers*, ch. III.; the quotation here is from Beman's translation, Chicago, 1901.

† Cf. Proj. Geom. § 19 (α).

hypothesis the series is such that there are two ways round from A to C, namely, one through B and one not through B. Let segm (ABC)

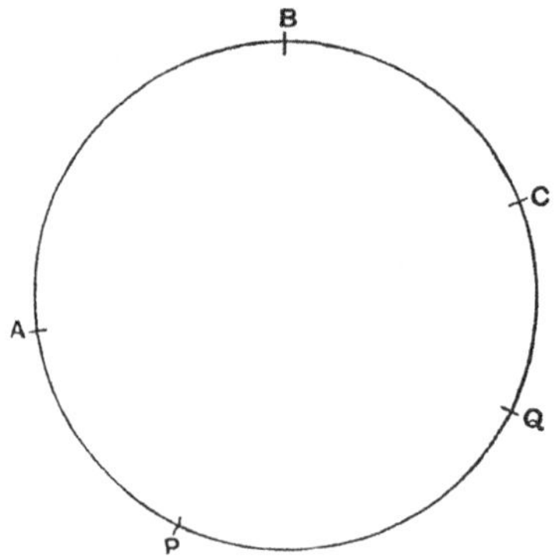

denote the points, excluding A and C, which are traversed from A to C through B, and let segm $(A\hat{B}C)$ denote the remaining points of the series. Again let a class u of the points of the series be called a segment of the series, when (1) there is a point B of the series which does not belong to u, and (2) if P and Q be any two points of u then segm $(P\hat{B}Q)$ belongs entirely to u.

Then the series possesses the Dedekind property if any segment such as u (which excludes more than one point of the series) must possess two boundary points, that is to say, if there must exist points A and C such that segm $(A\hat{B}C)$, with the possible exception of either or both of A and C, is identical with u. Here—as above—B is a point which does not belong to u.

Hence a sheaf of half-rays can also be considered as a closed compact series with the Dedekind property. This is made immediately evident by surrounding the vertex by a triangle in the plane of the sheaf. Then each half-ray of the sheaf intersects the boundary of the triangle in one and only one point. Also the order of the points on the boundary is the order of the corresponding half-rays of the sheaf. But the boundary of the triangle is a closed series with the Dedekind property.

10. By the aid of the Dedekind axiom and of the preceding axioms, it can now be proved that, if l be any line and A be any point, not incident in l, then in the plane Al at least one line can be drawn through A, which does not intersect l.

For take any point B on l, and let p and q be the two supplementary half-rays of l which emanate from B. Consider the sheaf of half-rays,

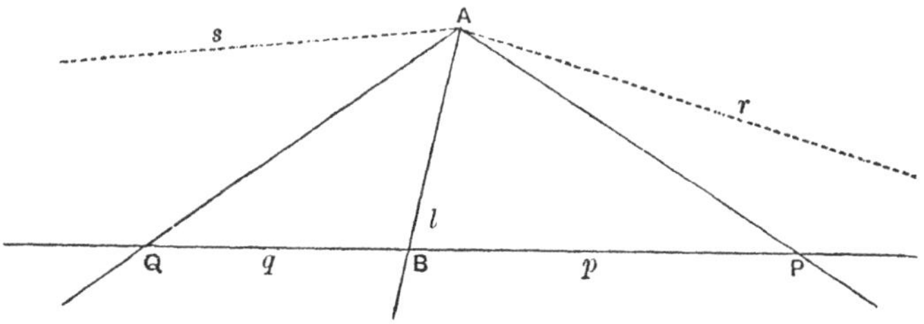

vertex A, in the plane Al. Some of these half-rays intersect p and some intersect q, and these classes are mutually exclusive.

Also, from the Dedekind property, there exist two semi-rays which are limits of the semi-rays intersecting p. AB is one of the semi-rays, let r be the other. Now the semi-ray p has no end-point. Hence r is not among the semi-rays intersecting p. Again by similar reasoning there is a semi-ray s which is the limit of the semi-rays intersecting q, and s does not intersect q.

Now first let r and s be not collinear, and let r' and s' be the half-rays supplementary to r and s respectively. Consider the set (α, say) of lines through A with one set of their half-rays between r and s', and therefore with their supplementary half-rays between r' and s. There are an infinite number of such lines. Now all half-rays emanating from A and lying between the half-ray AB and r intersect p, and no other half-rays from A intersect p. Similarly for the half-rays AB and s and q. Also if s' lie between the half-ray AB and r, then r' lies between the half-ray AB and s; and in this case every line of the set α intersects the line l twice, namely once for each of its pair of supplementary half-rays emanating from A. But this is impossible. Hence neither the supplement of r nor that of s can intersect l. Secondly if r and s are collinear, then the complete line formed by r and s cannot intersect l. For neither r nor s intersects l.

Thus taking any point A and any line l, the sheaf of lines, vertex A and in the plane Al, falls into two parts, namely the lines which intersect l, called the lines 'secant' to l, and the lines which do not intersect l, called the lines 'non-secant' to l. The non-secant lines of the sheaf may reduce to one line. The supposition that this is the case is the 'Euclidean Axiom.'

CHAPTER II.

THE ASSOCIATED PROJECTIVE SPACE.

11. We have now to establish the relation of Descriptive Geometry to an associated Projective Geometry. In a Projective Space let a 'convex region' be defined to be a region which (1) does not include the whole of any line, and (2) includes the whole of one of the two segments between any two points within it. It is easy to prove that such regions exist. For remembering* that we can employ the ordinary theory of homogeneous coordinates, the surface

$$x^2 + y^2 + z^2 - u^2 = 0$$

is well known to enclose such a region. Let a quadric enclosing a convex region be called a 'convex quadric.' Again in two dimensions let A, B, C be any three non-collinear points, and let P be any point not collinear with any two of them. Let AP meet the line BC in L, BP meet CA in M, CP meet AB in N.

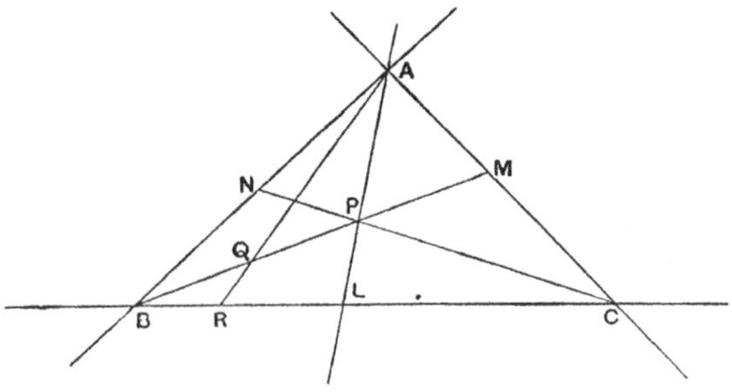

Define the triangular region (ABC/P) to be the set of points formed by the collection of segments such as segm (AQR)†, where Q is any point on segm (BPM), and R is the point where the line

* Cf. Proj. Geom. §§ 37 and 42.

† Cf. Proj. Geom. § 13.

AQ intersects BC. The points A, B, C can be interchanged in this definition without altering the region obtained.

Similarly in three dimensions, if A, B, C, D are the vertices of a non-degenerate tetrahedron, and P be any point not on any of the planes, ABC, BCD, etc., the 'tetrahedral region' $(ABCD/P)$ can be similarly defined. From the ordinary theory of homogeneous coordinates, it is well known that a triangular region in two dimensions, and a tetrahedral region in three dimensions, are both convex regions.

Again the triangular region (ABC/P) considered above has as its 'boundary' the segments (BLC), (CMA), (ANB), together with the points A, B, C. Also considering the tetrahedral region $(ABCD/P)$, let AP intersect BCD in L, BP intersect CDA in M, CP intersect DAB in N, DP intersect ABC in O. Then the 'boundary' of the tetrahedral region $(ABCD/P)$ consists of the triangular regions (BCD/L), (CDA/M), (DAB/N), (ABC/O), together with the boundaries of these triangular regions.

It is now a well-known result from the use of coordinates that in two dimensions any line through a point in a triangular region cuts the boundary in two points only; and that in three dimensions any line through a point in a tetrahedral region cuts the boundary in two points only.

12. Now consider a convex region, let it be either the region within a convex quadric, or a tetrahedral region. Call the points within it 'Descriptive points'; and call the portions of lines within it 'Descriptive lines.' The projective order of points on a line becomes an open order for Descriptive points on Descriptive lines. Then by the use of coordinate Geometry it is easy to prove that all the Descriptive axioms of the present tract, either in Peano's form or in

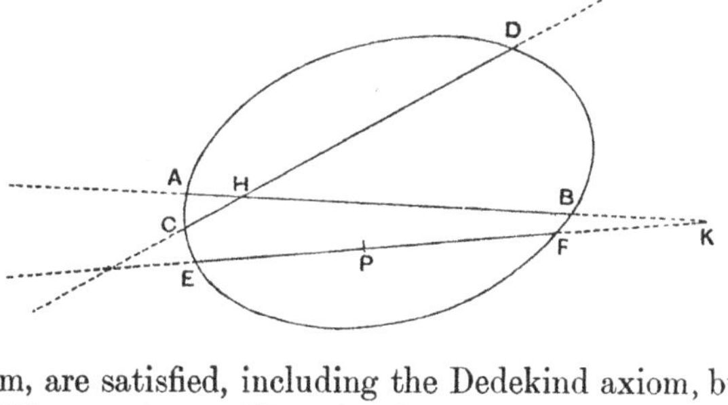

Veblen's form, are satisfied, including the Dedekind axiom, but excluding the Euclidean axiom. Thus in the figure the lines AB and CD

intersect at a point H in Descriptive Space; but the lines AB and EF do not intersect in Descriptive Space, since K lies outside it. Also it is evident that through any point P an infinite number of lines can be drawn, coplanar with AB, and not intersecting it in Descriptive Space.

13. The previous article (§ 12) proves* the existence theorem for Descriptive Space with the negation of the Euclidean axiom; in other words, it proves the independence of the Euclidean axiom.

The existence theorem for Descriptive Space with the Euclidean axiom is immediately proved by considering the region of Projective Space found by excluding all the points on one projective plane. The region is convex according to the above definition; also all the Descriptive axioms, together with the Dedekind axiom and the Euclidean axiom, hold for it†.

14. The independence of the Dedekind axiom of the other axioms, combined with the negation of the Euclidean axiom, is proved by considering, as in § 12, Descriptive Space to be a tetrahedral region in Projective Space, but confining ourselves to the points whose coordinates are algebraic numbers‡, as in the corresponding proof for Projective Geometry.

The independence of the Dedekind axiom of the other axioms, combined with the Euclidean axiom, is similarly proved by considering Descriptive Space to be the region in Projective Space found by

* Cf. Proj. Geom. § 43.

† In the later Greek period, and during the seventeenth and eighteenth centuries, the discussion of the foundations of Geometry was almost entirely confined to attempts to prove the Euclidean axiom. The explicit recognitions of its independence by Lobatschefskij (1828), and by J. Bolyai (1832) laid the foundation of the existing theories of non-Euclidean Geometry. For the literature of the whole question cf. Stäckel and Engel, *Die Theorie der Parallellinien von Euklid bis auf Gauss*, Leipzig, 1895, and also their *Urkunden zur Geschichte der Nichteuklidischen Geometrie, I. Lobatschefskij*, Leipzig, 1898.

‡ Cf. Proj. Geom. § 43 (a). An oversight in this proof may be here corrected. The proof, as printed, proceeds by considering only points with rational coordinates. But then a difficulty arises as to the theory of segments given in Chapter IV. of Proj. Geom. For it is necessary that the real double points of a hyperbolic involution should belong to the points considered. But these double points are given by a quadratic equation. Thus algebraic numbers (*i.e.* numbers which can occur as the roots of equations with integral coefficients) should be substituted for rational numbers. The proof proceeds without other alteration. I am indebted to Mr G. G. Berry of the Bodleian Library for this correction.

excluding a particular plane; and further, as before, we confine our consideration to the points whose coordinates are algebraic numbers.

15. It has been proved in §§ 12 and 13 that a convex region of a Projective Space is a Descriptive Space. The converse problem has now to be considered in this and in the next chapter; namely, given a Descriptive Space, to construct a Projective Space of which the Descriptive Space is part. This effects a very considerable simplification in the investigation of the properties of Descriptive Space owing to the superior generality of the analogous properties of Projective Space. Thus a Projective Space affords a complete interpretation of all the entities indicated in coordinate geometry. It is in order to gain this simplification that the 'plane at infinity' is introduced into ordinary Euclidean Geometry. We have in effect to seek the logical justification for this procedure by indicating the exact nature of the entities which are vaguely defined as the 'points at infinity'; and the procedure is extended by shewing that it is not necessarily connected with the assumption of the Euclidean axiom. This investigation is the Theory of Ideal Points*, or of the generation of 'Proper and Improper Projective Points' in Descriptive Geometry. The Euclidean axiom will not be assumed except when it is explicitly introduced. The remainder of this chapter will be occupied with the general theorems which are required for the investigation.

16. If A be any point and l be any line not containing A, then the plane Al divides the bundle of half-rays emanating from A into three sets, (1) the half-rays in the plane Al, (2) the half-rays on one side of the plane, (3) the half-rays on the other side of the plane. The sets (2) and (3) are formed of half-rays supplementary one to the other.

Lemma. With the above notation, it is possible to find a plane through the line l and intersecting any finite number of the half-rays either of set (2) or of set (3).

For let $a_1, \dots a_n$ be n half-rays of one of the two sets. Let B_1 be any point on a_1, and B_2 be any point on a_2. Then either the plane

* Originally suggested by Klein (extending an earlier suggestion of von Staudt), *Math. Annal.* vols. IV. and VI., 1871 and 1872; first worked out in detail by Pasch, *loc. cit.*, §§ 6—9. In the text I have followed very closely a simplification of the argument given by R. Bonola, *Sulla Introduzione degli Enti Improprii in Geometria Projectiva*, Giornale di Matematiche, vol. XXXVIII., 1900.

B_1l lies between the planes B_2l and Al, or the plane B_2l lies between the planes B_1l and Al, or the planes B_1l and B_2l are identical. But in either of the first two cases the intermediate plane intersects both semi-rays a_1 and a_2. Hence a plane is found through l, intersecting both a_1 and a_2. Call it the plane $B_2'l$. Again take any point B_3 on a_3; and the same argument shews that at least one of $B_2'l$ and B_3l intersects a_1, a_2, and a_3. Proceeding in this way, a plane is finally found which intersects each of the n semi-rays.

17. Desargues' Perspective theorems* can be enunciated in the following modified forms :

(1) If two coplanar triangles ABC and $A'B'C'$ are such that the lines AA', BB', CC' are concurrent in a point O, then the three inter-

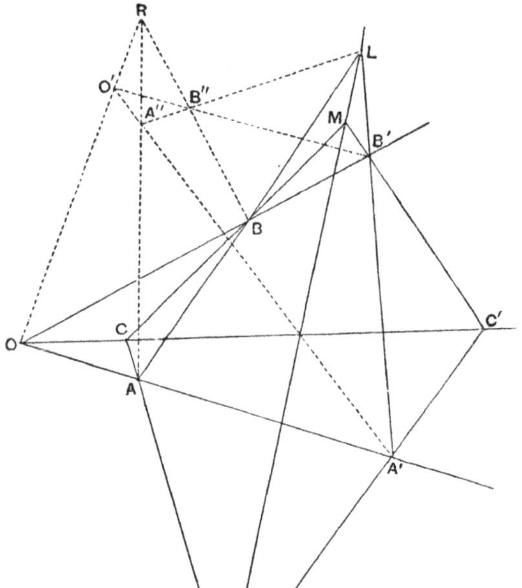

sections of BC and $B'C'$, CA and $C'A'$, AB and $A'B'$, if they exist, are collinear.

* Cf. Proj. Geom. § 7.

(2) If the pairs of homologous sides of the two coplanar triangles ABC and $A'B'C'$, namely, BC and $B'C'$, CA and $C'A'$, AB and $A'B'$, intersect in three collinear points, then the lines AA', BB', CC', if any two intersect, are concurrent in the same point.

Considering the first proposition let AB and $A'B'$ intersect in L, BC and $B'C'$ in M, CA and $C'A'$ in N. Now it is not possible both for A' to lie on the segment OA, and for A to lie on the segment OA'. Assume that A' does not lie on the segment OA. Let R be any point external to the given plane (α, say). Now by the lemma of § 16, it is possible to find a plane through LM, lying between the planes LMR and LMA (*i.e.* the plane α), and intersecting the three lines RA, RB, RC, say, in the points A'', B'', C'' (in the figure C'' is not shewn). Then evidently A'' must lie in the segment RA. Hence $A'A''$, since A' does not lie in the segment OA, must intersect OR in the segment OR. Thus the intersection of the lines $A'A''$ and OR is secured. Let it be the point O'. Again the lines $O'A'$ and $O'B'$ are the projections from R on the plane $A'O'B'$ of the lines OA' and OB'. Now $A''B''$ passes through L. Hence B'' lies on the plane $A'A''B'$, *i.e.* on the plane $OA'B'$. Hence B'' is on the projection of the line OB' on the plane $O'A'B'$, *i.e.* B'' lies on $O'B'$. Thus $B'B''$ passes through O'.

Reasoning in exactly the same way for BC and $B'C'$, it follows that $C'C''$ passes through O'. The same figure has now been constructed as in the proof of the corresponding theorem for Projective Geometry*. Accordingly the theorem follows by the same reasoning.

In order to demonstrate the converse theorem, we proceed exactly as above, except that, L, M, N are now assumed to be collinear, O is the point of intersection of AA' and BB'. Then the same construction is made as before, and it is successively proved by similar reasoning that every pair of the lines $A'A''$, $B'B''$, $C'C''$ intersect. But the lines are not coplanar. Hence they intersect in the same point O'. But O' must lie on RO. Thus CC' passes through O.

Corollary. The enunciation of the first theorem can be modified by removing the assumption that AC and $A'C'$ intersect, but by adding the assumption that AC intersects LM.

18. A trihedron is the figure formed by three lines concurrent in the same point, and not all coplanar. The three lines form the edges of the trihedron; the three planes containing the lines, two by

* Cf. Proj. Geom. § 7.

two, form the faces of the trihedron; the point of concurrence of the three edges is the vertex of the trihedron.

It follows (cf. § 6) that, if two trihedrons have the same vertex, any two faces, one from each trihedron, must intersect in a line through the vertex; also that any two planes each containing two edges, one edge from each trihedron, must intersect in a line through the vertex.

Desargues' theorems can be applied to two trihedrons with the same vertex; only in this case, as in Projective Geometry, there are no exceptional cases depending on non-intersection.

The enunciations are as follows:

(1) If a, b, c and a', b', c' are the edges of two trihedrons with the same vertex, such that the planes containing a and a', b and b', c and c', are concurrent in a line s (*i.e.* belong to the same sheaf), then the three intersections of the planes bc and $b'c'$, ca and $c'a'$, ab and $a'b'$ are coplanar.

(2) If a, b, c and a', b', c' are the edges of two trihedrons with the same vertex, such that the three intersections of the planes bc and $b'c'$, ca and $c'a'$, ab and $a'b'$, are coplanar, then the three planes containing a and a', b and b', c and c' belong to the same sheaf.

These propositions immediately follow from the case of triangles by noticing that, by the lemma of § 16, the six edges of the trihedrons can be cut by a plane, not through the vertex. Hence by the previous remarks on trihedrons, Desarguesian triangles are obtained without the exceptions due to non-intersection.

19. The two theorems of the present and next articles are the central theorems of the whole theory of Ideal Points.

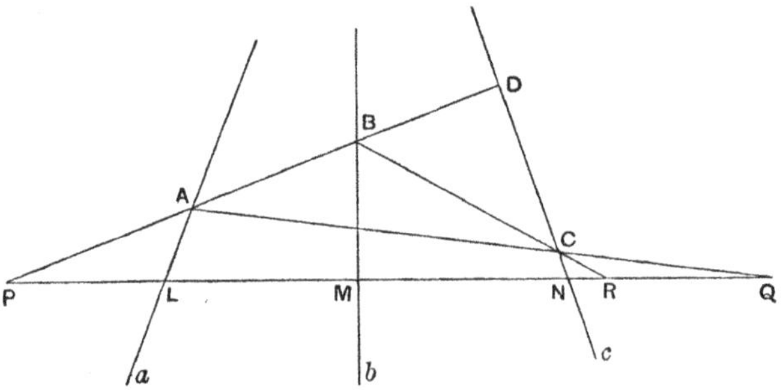

If the lines a, b, c are the intersections of three planes α, β, γ of a sheaf with a plane π, not belonging to the sheaf, and if O be any point not incident in π, then the three planes Oa, Ob, Oc belong to one sheaf.

If the axis (r) of the sheaf intersect the plane π in a point S, then the three lines a, b, c pass through S, and the line OS is evidently contained in Oa, Ob, Oc, and thus forms the axis of the new sheaf.

Consider now the case when the axis (r) of the sheaf does not intersect π. Then a, b, c are not concurrent, and no two of them intersect. Hence one of the three (b, say) must lie between [*i.e.* any segment, joining a point on a and a point on c, intersects b] the other two. Take any two points L and N on a and c respectively, then the segment LN intersects b in a point M. Take two other points P and Q on LN so that we have the order P, L, M, N, Q. Take any point D on c; then the segment PD must intersect a and b in two points A and B respectively; and the segment AQ must intersect the segment DN in a point C. Then the line BC must intersect the segment NQ in a point R. Thus a triangle ABC has been formed, whose vertices lie on a, b, c, and whose sides AB, AC, BC pass through P, Q, R respectively.

By taking another point D' on c, another triangle $A'B'C'$ can be similarly formed, whose vertices lie on a, b, c, and whose sides $A'B'$ and $A'C'$ pass through P and Q respectively.

We have first to shew that $B'C'$ passes through R. For taking any point T on the axis (r) of the sheaf, the lines TA, TB, TC form the edges of one trihedron, and the lines TA', TB', TC' form the edges of another trihedron with the same vertex.

Also the planes TAA', TBB', TCC' belong to the same sheaf. Hence the three intersections of the pairs of planes TAB and $TA'B'$, TAC and $TA'C'$, TBC and $TB'C'$ are coplanar; hence they lie in the plane TPQ. Hence $B'C'$ passes through R.

Now considering the two trihedrons with edges OA, OB, OC, and OA', OB', OC', the intersections of the pairs of faces OAB and $OA'B'$, OAC and $OA'C'$, OBC and $OB'C'$, are respectively OP, OQ, OR; and these are coplanar. Hence by the converse part of Desargues' theorem for trihedrons, the planes OAA', OBB', OCC' belong to the same sheaf. Hence Oa, Ob, Oc belong to the same sheaf (*i.e.* have a common line of intersection).

20. If any two of the lines a, b, c are coplanar, but the three lines are not coplanar, and similarly for the lines a, b, d, then c and d are coplanar.

If a and b intersect, the theorem is evident; for a, b, c are concurrent, and a, b, d are concurrent. Hence c and d are concurrent.

Assume that a and b do not intersect. Then it is easy to prove that no two of the lines intersect. It follows that no one of the lines c, b, d can intersect any of the planes ab, ac, ad in which it does not lie.

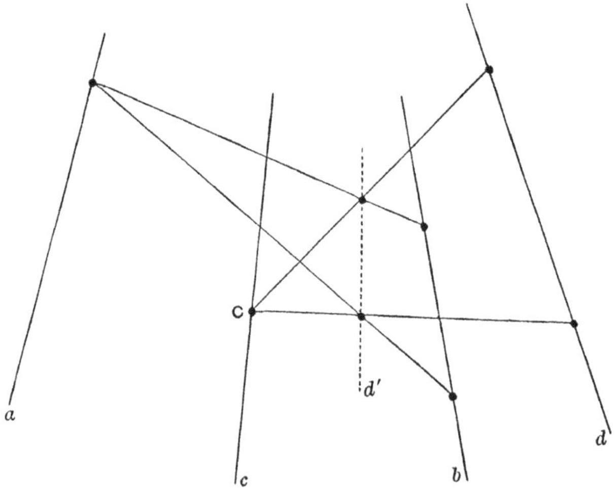

Fig. 1.

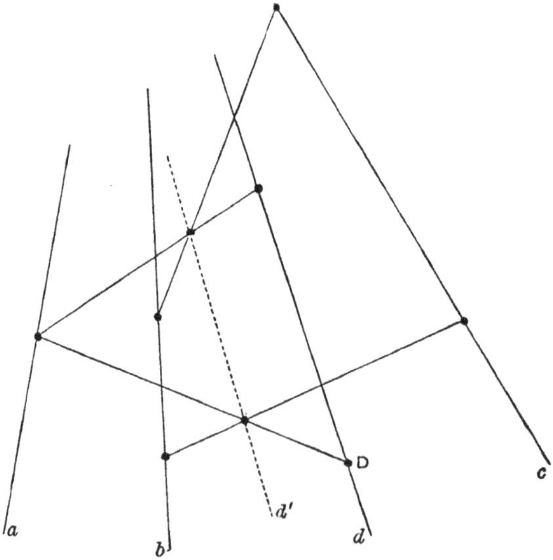

Fig. 2.

Hence it follows that either c and d lie on opposite sides of the plane ab, or d and b on opposite sides of the plane ac, or b and c on opposite sides of the plane ad.

First, let c and d lie on opposite sides of the plane ab (cf. fig. 1). Take any point C on c. Then the plane Cd must intersect the plane ab in a line, d', say. Then the lines a, b, d' are the intersections of the three planes da, db, dC with the plane ab; and these three planes belong to the same sheaf. Hence (cf. § 19) the three planes through the lines a, b, d' respectively and through any point not on ab belong to the same sheaf. But C is such a point. Hence the planes Ca, Cb, Cd' belong to the same sheaf. But c is the common line of Ca and Cb. Hence Cd' contains the line c. Hence c and d are coplanar.

Secondly, let the plane ad lie between b and c. Then the plane bc must intersect the plane ad in some line, d' say. Thus the three lines b, d', c are the intersections of the three planes ab, ad, ac with the plane bc. These three planes belong to the same sheaf. Hence (cf. § 19), if D is any point on d, not on bc, the planes Db, Dc, Dd' belong to the same sheaf. But Db and Dd' intersect in the line d; hence Dc passes through the line d. Thus c and d are coplanar.

Thirdly, let the plane ac lie between b and d. Then the proof is as in the second case, interchanging c and d.

CHAPTER III.

IDEAL POINTS.

21. *Definition.* An 'Associated* Projective Point,' or an 'Ideal Point,' is the class of lines which is composed of two coplanar lines, a and b, say, and of the lines formed by the intersections of pairs of distinct planes through a and b respectively, and of the lines in the plane ab which are coplanar with any of the lines of the projective point not lying in the plane ab.

It follows immediately from § 20 that the lines forming a projective point are two by two coplanar; and further that (with the notation of the definition) the lines of the projective point lying in the plane ab are the lines in ab coplanar with any *one* of the lines of the projective point not lying in ab.

Definition. A projective point is termed 'proper,' if the lines composing it intersect. Their point of intersection will be called the 'vertex' of the point.

Thus a proper projective point is simply a bundle of lines, and every bundle is a proper projective point.

Definition. A projective point is termed 'improper,' if the lines composing it do not intersect.

It is proved (cf. §§ 24—30) that Projective Geometry holds good of projective points as thus defined, when a fitting definition has been given of a 'projective line.'

Definition. A projective point will be said to be 'coherent with' a plane, if any of the lines composing it lie in the plane.

Definition. A 'projective line' is the class of those projective points which are coherent with two given planes. If the planes

* The word 'Associated' will usually be omitted.

intersect, the projective line is called 'proper'; and the line of intersection is the 'axis.' If the planes do not intersect, the projective line is called 'improper.'

Since Projective Geometry has been developed* from the two fundamental ideas of 'point' and 'straight line,' the other definitions of projective elements must simply be those which have been given in considering Projective Geometry. Thus† a projective plane is the class of those projective (ideal) points, which lie on any projective line joining any given projective point A to any projective point on any given projective line not possessing the given projective point A.

Definition. If a projective plane possesses any proper projective points, it will be called a 'proper projective plane.' Otherwise it is an 'improper projective plane.'

The vertices of all the proper projective points on a proper projective plane will be seen to form a plane (cf. § 26 (α)).

Definition. A proper projective point and its vertex are said to be 'associated,' so likewise are a proper projective line and its axis, and also a proper projective plane and the plane constituted by the vertices of its proper projective points.

22. Since any two lines belonging to a projective point are coplanar, it easily follows that any two lines of the projective point can be used in place of the two special lines (a and b) used in the definition (cf. § 21). Hence it can easily be proved that any plane, containing one line of a projective point, contains an infinite number of such lines. In other words, if a projective point is coherent with a plane, an infinite number of the lines of the projective point lie in the plane. In fact it follows that, through each point of the plane, one line passes which belongs to the projective point.

23. If three projective points are incident in the same projective line, then with any plane, with which two of the projective points cohere, the third projective point also coheres.

First, if the three projective points are proper, the theorem is immediately evident.

Secondly, let two of the projective points, M and N, say, be proper, and let the third projective point, L, say, be improper. Let

* Cf. Proj. Geom.

† Cf. Proj. Geom. § 4.

the projective line possessing L, M, N be defined by the two planes π and π' (cf. Definition of § 21). Then the three projective points cohere with π and π'. Let π'' be any third plane with which two of the three,

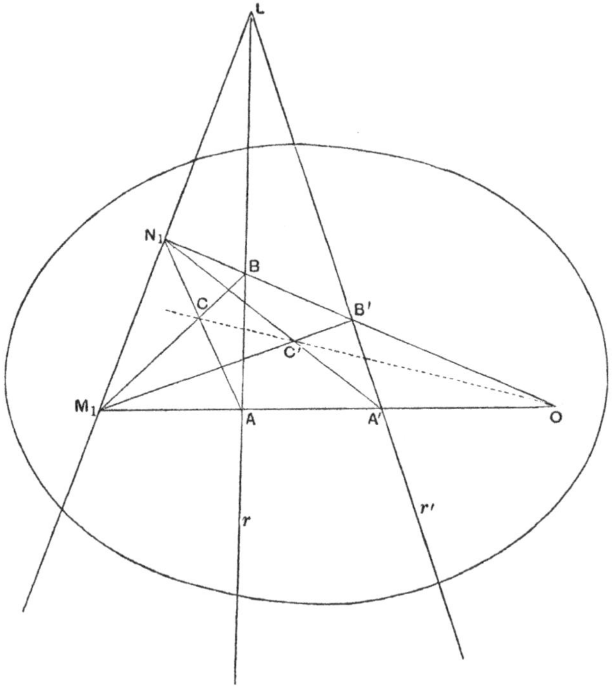

Fig. 1.

L, M, and N, cohere. Let M_1 and N_1 be the vertices of the proper points, M and N. Then M_1N_1 is a line in the plane π (cf. fig.* 1); also the line M_1N_1 belongs to L. Again (cf. § 22) another line r exists in π belonging to L; and M_1 and N_1 must lie on the same side of r. Let r' be any line in π belonging to L, and on the opposite side of the line r to M_1 and N_1; such a line exists (cf. § 22). Let O be any point of π on the side of r' remote from M_1 and N_1. Then the segment OM_1 intersects r and r', in A and A', say; and the

* Note in drawing an illustrative figure, it is convenient to make the assumption of § 12, and to consider Descriptive space as a convex region in a larger Projective Space. This region is marked off by an oval curve in the figure, and an ideal point, such as L, is a point outside the oval. Note that the existence of L, as an analogous entity to M_1 and N_1, must not be assumed in the present reasoning.

segment ON_1 intersects r and r', in B and B', say. The segments AN_1 and BM_1 intersect, in C, say; the segments $B'M_1$ and $A'N_1$ intersect, in C', say. Now project from any point O' in π', and two trihedrons are formed, namely $O'A$, $O'B$, $O'C$, and $O'A'$, $O'B'$, $O'C'$, with the same vertex O'. Also the homologous faces intersect in the three coplanar lines $O'L$, $O'M_1$, $O'N_1$. Hence the three planes $O'AA'$, $O'BB'$, $O'CC'$ are concurrent in a line. Hence the plane $O'CC'$ contains the line $O'O$. Therefore CC' passes through O. Again project from any point O'' in π'', and consider the trihedrons $O''A$, $O''B$, $O''C$, and $O''A'$, $O''B'$, $O''C'$. Then the planes $O''AA'$, $O''BB'$, $O''CC'$ are concurrent in the same line $O''O$. Thus the three lines $O''L$, $O''M_1$, $O''N_1$ are coplanar. Hence if two of them lie in π'', the third must do so also. Hence if two of L, M, N cohere with π'', the third also does so.

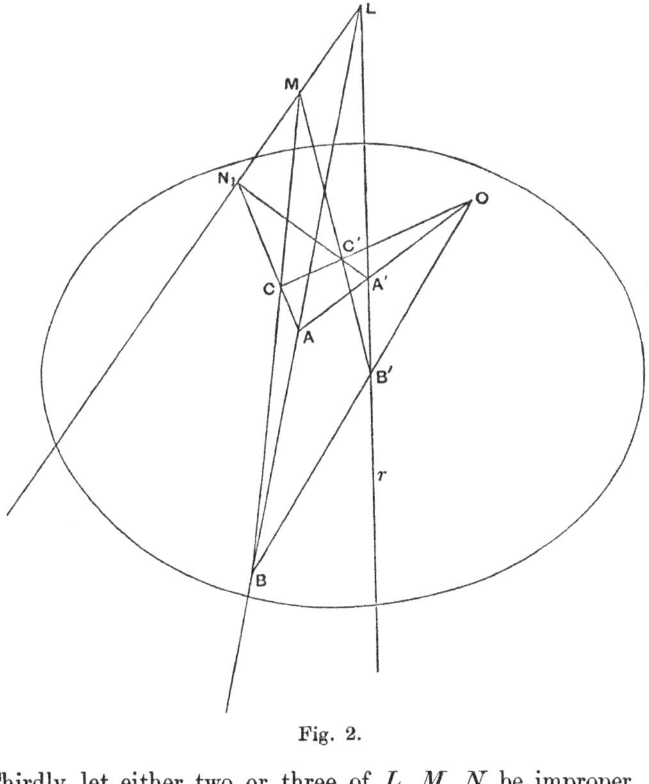

Fig. 2.

Thirdly, let either two or three of L, M, N be improper. Thus let L and M be certainly improper (cf. figs. 2 and 3). In the plane π

form a triangle ABC, such that its sides AB, BC, CA belong to L, M, N respectively. Thus if N is a proper point, CA passes through N_1, the vertex of N: also since the lines BC and AB do not intersect the line N_1ML, the points A, B, C lie on the same side of this line (cf. fig. 2): also since the lines BC and BA do not intersect the line N_1ML, either C lies on the same side of the line AB as N_1, or A lies on the same side of the line BC as N_1. Assume that C lies on the same side of AB as N_1 (cf. fig. 2). The rest of the proof for figures 2 and 3 is now identical. In the plane π, let r be any line

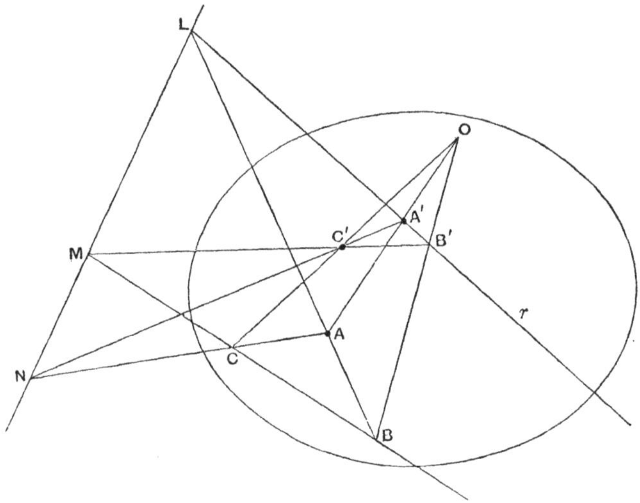

Fig. 3.

belonging to L, on the side of AB remote from C. In the plane π, take any point O on the side of r remote from C. Then the segments OA and OB intersect r, say in A' and B'. Also the line $A'N$ intersects the segment OA, and does not intersect the segment AC; hence it must intersect the segment OC, say in C'. Then, by projecting from any point O' in the plane π' and by similar reasoning to that in the second case, it is proved that the line $B'C'$ belongs to M. Then, as in the second case, by projecting from any point O'' in π'', it follows that, if any two of the projective points L, M, N cohere with π'', so also does the third.

24. (α) It follows from § 23 that any two planes, with which both of two given projective points cohere, define the same projective line as any other pair of such planes. Hence two projective points determine not more than one projective line.

(β) Two projective points determine at least one projective line. For if the points are proper, this is immediately evident. But in any case let the projective points be A and B, and let O be any point. There are at least two lines, a_1 and a_2, which are members of A and such that the plane a_1a_2 does not contain O. Then the planes Oa_1 and Oa_2 intersect in a line which passes through O and is a member of A. Hence through any point there passes a line which is a member of a projective point. Hence through O there are lines belonging to A and B respectively. But these lines determine a plane, with which A and B both cohere. Similarly a second such plane can be determined. Hence there is a projective line possessing both A and B.

25. The Axioms of Projective Geometry* can now be seen to be true for the 'Projective Elements' as thus defined. Thus we have the following theorems corresponding to those axioms of the previous tract, of which the numbers are enumerated in the initial brackets.

(I, II, III.) There is a class of Projective Points, possessing at least two members.

(IV, V, VI, VII, VIII.) If A and B are Projective Points, there is a definite projective line AB, which (1) is a class of projective points, and (2) is the same as the projective line BA, and (3) possesses A and B, and (4) possesses at least one projective point distinct from A and B.

Note that two improper projective points may possess no common line.

(IX and X.) If A and B are projective points, and C is a projective point belonging to the projective line AB, and is not identical with A, then (1) B belongs to the projective line AC, and (2) the projective line AC is contained in the projective line AB.

(XI.) If A and B are distinct projective points, there exists at least one projective point not belonging to the projective line AB.

26. Before considering the proof of the 'axioms' of the projective plane†, some further propositions are required.

(a) Since a line exists through any given point and belonging to any given projective point, it easily follows that the set of projective

* Cf. Proj. Geom. §§ 4, 7, 8, 14.

† Cf. Proj. Geom., Axioms XII, XIII, XIV.

points cohering with a plane form a proper projective plane; and that conversely, any proper projective plane is the set of projective points cohering with some plane.

(β) Any projective line intersects any given proper projective plane. For through the vertex of any proper projective point on the projective plane, a plane passes with which every point of the projective line coheres (cf. § 23). This plane intersects the plane associated with the projective plane in a line. Two such planes can be found. The two lines in the plane associated with the projective plane define a projective point which lies both in the projective line and the projective plane.

(γ) Two projective lines in a proper projective plane necessarily intersect.

For let m and n be the projective lines and α be the proper projective plane, and α_1 its associated plane. Take any point O outside α_1. Then two planes Om and On exist, with which respectively all projective points of m and n cohere. These planes intersect in a line through O, l, say. Let A be any point in α_1. The plane Al intersects α_1 in a line, l', say. The two lines l and l' define a projective point which lies in both the projective lines m and n.

(δ) Desargues' Theorem holds for triangles formed by projective lines and projective points in a proper projective plane.

By (γ) immediately above, no exception arises from non-intersection. Then by taking a point external to the associated plane, two trihedrons can be formed for which the theorem holds. Hence the theorem holds for the proper projective plane.

(ϵ) The projections upon a proper projective plane of three projective points belonging to the same projective line also belong to a projective line.

The theorem is immediately evident, if the centre of projection, or if any one of the three projective points, is proper. Assume that all the projective points are improper. Let L, M, N be the three projective points, and S the projective point which is the centre of projection. Let π be the proper projective plane on to which L, M, N are to be projected. Let α be any plane with which L, M, N all cohere. On α construct figure 3 of § 23. Project (remembering (β) above) the whole figure of associated projective points from S on to the plane π. Then by the first case of the present theorem, all collinear groups of projective points which possess a proper projective point are projected into collinear groups. Let A, B, ... M, N be projected into A_1, B_1, ... M_1, N_1.

Thus, in the plane π, two homological triangles $A_1B_1C_1$ and $A_1'B_1'C_1'$ are obtained, A_1A_1', B_1B_1', C_1C_1' being concurrent in O_1; also B_1A_1 and $B_1'A_1'$, B_1C_1 and $B_1'C_1'$, A_1C_1 and $A_1'C_1'$ are concurrent respectively in L_1, M_1, N_1. Hence, by (δ) above, L_1, M_1, N_1 belong to the same projective line.

27. The next group of propositions correspond to the three axioms concerning the projective plane.

(XII.) If A, B, C are three projective points, which do not belong to the same projective line, and A' belongs to the projective line BC, and B' to the projective line CA, then the projective lines AA' and BB' possess a projective point in common.

If the projective plane ABC is proper, the theorem follows from § 26 (γ). If the projective plane ABC is improper, consider any plane with which all the projective points of the projective line BB' cohere.

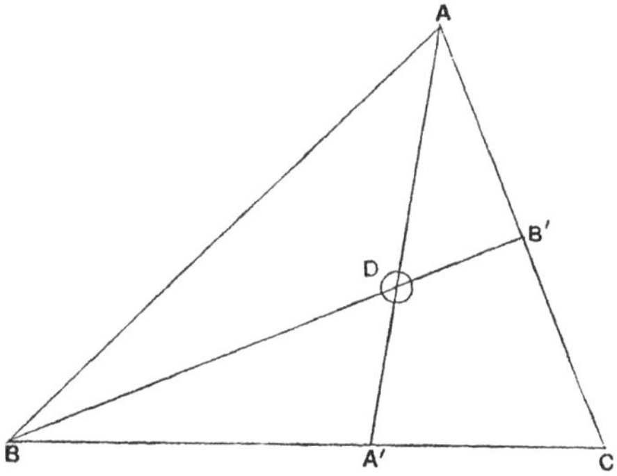

Such planes exist. Thus the associated projective plane of such a plane is a proper projective plane containing the line BB'. But by § 26 (β) the projective line AA' intersects this proper projective plane, in the projective point D, say. Also by § 26 (ϵ) the projections of B, A', C from A on to this proper projective plane belong to the same projective line. Hence D belongs to BB'. Thus AA' and BB' intersect.

(XIII.) If A, B, C are three projective points, not belonging to the same projective line, then there exists a projective point not belonging to the projective plane ABC.

This follows immediately from Peano's Axiom XV given in § 6 above.

28. The theory of Harmonic Ranges must now be considered.

Let A, B be any two points, C a point in the segment AB. Take F any point outside the line AB, and H any point on the segment FC, and let EG be as in figure 1. Then the point D, if it exist, is

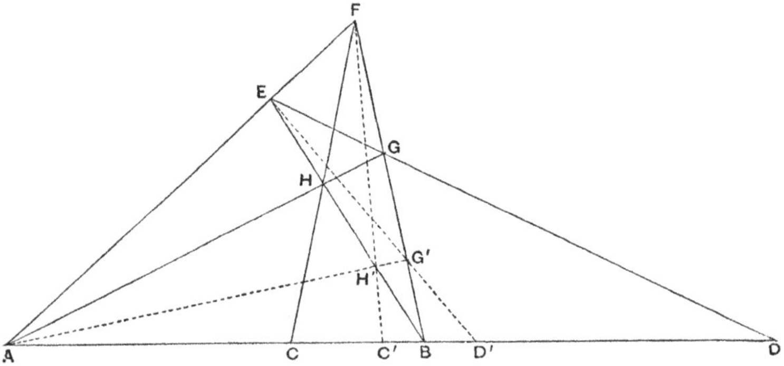

Fig. 1.

the harmonic conjugate of C with respect to A and B. By considering the associated projective points and the associated projective lines, the requisite harmonic conjugate (as a projective point) always exists. Thus, on the basis of the axioms of Projective Geometry already proved, the proof for the uniqueness of the harmonic conjugate in the associated projective geometry holds good*. Thus in the original Descriptive Geometry, the harmonic conjugate, if it exist, is unique.

Furthermore, since H is on the segment FC, E and G are respectively on the segments AF and FB. Hence D, if it exist, cannot lie on the segment AB. Conversely, if D is any point on the line AB, say on the side of B remote from A, take E any point on the segment AF, then DE must intersect the segment BF. Hence AG and BE must intersect in H, on the segments AG and BE. Therefore FH intersects the segment AB. Thus the harmonic conjugate with respect to A and B of any point on the line AB, not on the segment AB, must exist, and lies on the segment AB.

Furthermore, if D lies on the side of B remote from A, and C' lies on the segment BC, let FC' and EB intersect in H'; and AH' intersect the segment FB in G'. Then since C' lies in the segment BC, H' lies in the segment BH, and G' lies in the segment BG. Hence D' exists and lies in the segment BD. Thus as C moves towards B, D also moves in the opposite direction towards B†.

* Cf. Proj. Geom. §§ 6 and 7. † Cf. Proj. Geom. § 17 (β).

Hence it is easily seen that the segment AB is divided into three parts by reference to the harmonic conjugates of points in it with respect to A and B. The part formed by the segment AK_1 (cf. fig. 2),

Fig. 2.

exclusive of A and K_1, contains the points whose harmonic conjugates lie on the side of A remote from B; the segment BK_2, exclusive of B and K_2, contains the points whose harmonic conjugates lie on the side of B remote from A; the segment K_1K_2, inclusive of K_1 and K_2, contains the points for which the harmonic conjugates do not exist. It is not necessary that the points K_1 and K_2 be distinct. If they coincide, the segment K_1K_2, inclusive of K_1 and K_2, shrinks into a single point K. Thus in Euclidean Geometry the middle point of any segment AB is this degenerate portion of the segment.

It immediately follows that Fano's axiom* is satisfied for proper projective lines. Hence, remembering that the harmonic relation is projective†, we have:

(XIV.) If A and B are distinct projective points, and C is a projective point of the projective line AB, distinct from A and B, then the harmonic conjugate of C, with respect to A and B, is distinct from C.

Also the restriction to three dimensions follows at once from Peano's Axiom XVI of § 6, giving the same restriction for Descriptive Geometry. Hence we find:

(XV.) If α be any projective plane, and A be any projective point not lying in α, any projective point P lies on some line joining A to some projective point on α.

29. The order of the projective points on a projective line must now be considered.

If the projective line is proper, the order of the proper projective points on it will be defined to correspond to the order of the associated points. Thus (cf. fig. 2 of § 28) if the points marked in the figure are projective points, as C moves from A to K_1, excluding K_1, the projective point D, which is the harmonic conjugate to C with respect to A and B, moves from C through all the proper projective points on the

* Cf. Proj. Geom. § 8.

† Cf. Proj. Geom. § 9 (δ).

side of A remote from B; and as C moves from K_2, excluding K_2, to B, D moves towards B through all the proper projective points on the side of B remote from A.

Now let the order of the improper projective points be defined so as to make the above theorem hold generally: thus as C moves from K_1 to K_2, including K_1 and K_2, let the order of the improper projective points through which D moves be such that D passes continually in the same direction round the line from the proper projective points on the side of A remote from B to the proper projective points on the side of B remote from A.

Then by theorem (α) of § 17 of the Tract on Projective Geometry, the order as thus defined agrees with the order as defined in §§ 14 and 15 of that Tract. Also the order on the improper projective lines is obtained from the order on the proper projective lines by projection. Since the harmonic property is projective, the orders obtained thus by different projections must agree. Also from Peano's axioms of the segments of the Descriptive line given in § 3 above, it follows that the Projective axioms of order* are satisfied, namely:

(XVI.) If A, B, C are distinct projective points on the same projective line, and D is a projective point on segm $(A\hat{B}C)$†, not identical either with A or C, then D belongs to segm (BCA).

(XVII.) If A, B, C are distinct projective points on the same projective line, and D is a projective point belonging to both segm (BCA) and segm (CAB), then D cannot belong to segm (ABC).

(XVIII.) If A, B, C are distinct projective points on the same projective line, and D is a projective point, distinct from B, and belonging to segm (ABC) [which excludes A and C], and E belongs to segm (ADC), then E belongs to segm (ABC).

30. Finally, the Dedekind property‡ for the projective line follows immediately from its assumption for Descriptive Geometry (cf. § 9 above).

Thus all the axioms for Projective Geometry, including the axioms of order and the Dedekind property, are satisfied by the Projective Points and the Projective Lines. Furthermore the proper projective points evidently form a convex region in the projective space formed by the projective points. Also the geometry of this convex region of

* Cf. Proj. Geom. § 14.

† *i.e.* on the segment between A and C not possessing B, cf. Proj. Geom. § 13.

‡ Cf. Proj. Geom. § 19 (α).

proper projective points corresponds step by step with the geometry of the original descriptive space. Thus the geometry of descriptive space can always be investigated by considering it as a convex region in a projective space. This simply amounts to considering the associated proper projective points and adding thereto the improper projective points. A particular case arises when the Euclidean axiom (cf. § 10, above) is assumed. The improper projective points then lie on a single improper projective plane. Thus in Euclidean Geometry when the 'plane at infinity' is considered, the associated projective geometry has been introduced, and this plane is the single improper projective plane.

CHAPTER IV.

GENERAL THEORY OF CORRESPONDENCE.

31. In this chapter the general ideas of Correspondences, or Transformations, and of groups of transformations are explained, and thus the idea of continuous motion is led up to.

Consider any proposition respecting two entities p and q; let it be denoted by $\phi(p, q)$. The proposition may be varied by replacing p and q by two other entities, say u and v, so that the new proposition is $\phi(u, v)$. Thus we arrive at the notion of a constant form common to all the propositions of the type $\phi(x, y)$, where x and y are any two entities such that a significant proposition results when x and y replace p and q in $\phi(p, q)$. Note that a false proposition is significant; an insignificant proposition is not in truth a proposition at all, it is a sequence of ideas lacking the type of unity proper to a proposition.

The constant form of the proposition $\phi(x, y)$, as x and y vary, may be said to constitute a *relation* between x and y, in those special cases for which $\phi(x, y)$ is a true proposition. The order of x and y in respect to this relation represents the special roles of x and y respectively in the proposition $\phi(x, y)$. Thus if this relation is called R, 'x has the relation R to y,' or more briefly xRy, is equivalent to the proposition $\phi(x, y)$, however x and y be varied. It is evident that we might have considered the relation indicated by the proposition in such form that, if it be denoted by R', $yR'x$ represents $\phi(x, y)$. Then R and R' are called mutually converse relations. It is evident that each relation has one and only one corresponding converse relation.

When xRy holds, x is called the referent and y the relatum. A relation is said to be a one-one relation when to each referent there is only one relatum, and to each relatum there is only one referent. For example, if aRb and aRc both hold, where b and c are distinct entities, then the relation R is not one-one.

The class of all the referents in respect to a relation is called the domain of the relation, and the class of all relata is the converse domain. In mathematics a one-one relation is often spoken of as a transformation (or correspondence) of the members of its domain into (or with) the corresponding members of the converse domain. The correspondence is definite and reversible, and constitutes a rule by which we can pass from any member of one class to a corresponding definite member of the other class.

For example, the equation,

$$2x + 3y = 4,$$

constitutes a one-one relation of all real numbers, positive or negative, to the same class of real numbers. This brings out the fact that the domain and the converse domain can be identical.

Again, a projective relation between all the points on one line of projective space and all the points on another (or the same) line constitutes a one-one relation, or transformation, or correspondence, between the points of the two lines. Any one-one relation of which both the domain and the converse domain are each of them all the points of a projective space will be called a one-one point correspondence.

32. By reasoning* based upon the axioms of Projective Geometry, without reference to any idea of distance or of congruence, coordinates can be introduced, so that the ratios of four coordinates characterize each point, and the equation of a plane is a homogeneous equation of the first degree. Let X, Y, Z, U be the four coordinates of any point; then it will be more convenient for us to work with non-homogeneous coordinates found by putting x for X/U, y for Y/U, z for Z/U. Accordingly the actual values of x, y, z are, as usual, the coordinates characterizing a point. All points can thus be represented by finite values of x, y, z, except points on the plane, $U=0$. For these points some or all of x, y, and z are infinite. In order to deal with this plane either recourse must be had to the original homogeneous coördinates, or the limiting values of x to y to z must be considered as they become infinite.

The plane, $x=0$, is called the yz plane, the line, $y=0$, $z=0$, is called the axis of x, and the plane, $U=0$, is called the infinite plane.

* Cf. Proj. Geom. chs. VI. and VII.

When the fundamental tetrahedron is changed, the coordinates are changed according to the formula

$$X' = \sigma\{a_{11}X + a_{12}Y + a_{13}Z + a_{14}U\},$$
$$Y' = \sigma\{a_{21}X + a_{22}Y + a_{23}Z + a_{24}U\},$$
$$Z' = \sigma\{a_{31}X + a_{32}Y + a_{33}Z + a_{34}U\},$$
$$U' = \sigma\{a_{1}X + a_{2}Y + a_{3}Z + a_{4}U\}.$$

Hence the non-homogeneous coordinates are transformed by the formula

$$x' = (a_{11}x + a_{12}y + a_{13}z + a_{14})/(a_1x + a_2y + a_3z + a_4),$$
$$y' = (a_{21}x + a_{22}y + a_{23}z + a_{24})/(a_1x + a_2y + a_3z + a_4),$$
$$z' = (a_{31}x + a_{32}y + a_{33}z + a_{34})/(a_1x + a_2y + a_3z + a_4).$$

But if the infinite plane is the same in both cases, the formula for transformation becomes

$$x' = a_{11}x + a_{12}y + a_{13}z + a_{14},$$

with two similar equations.

33. A one-one point correspondence can be characterized by formulæ giving the coordinates of any relatum in terms of those of the corresponding referent. It must be remembered that every point is both a relatum and a referent. Let the correspondence under consideration be called T; then the coordinates of the relatum of any point x, y, z will be written Tx, Ty, Tz. Thus we have,

$$Tx = \phi_1(x, y, z),\quad Ty = \phi_2(x, y, z),\quad Tz = \phi_3(x, y, z),$$

where the functions ϕ_1, ϕ_2, ϕ_3 are defined for every point of space and are single-valued. Furthermore, since the correspondence, being one-one, is reversible, it must be possible to solve these equations for x, y, and z, obtaining

$$x = \psi_1(Tx, Ty, Tz),\quad y = \psi_2(Tx, Ty, Tz),\quad z = \psi_3(Tx, Ty, Tz).$$

Let this converse relation be written T_1, and let the coordinates of the relatum of any point x, y, z be written T_1x, T_1y, T_1z. Then

$$x = T_1Tx = TT_1x,\quad y = T_1Ty = TT_1y,\quad z = T_1Tz = TT_1z.$$

Then, remembering that by properly choosing x, y, z we can take Tx, Ty, Tz to be any point of space, we find

$$T_1x = \psi_1(x, y, z),\quad T_1y = \psi_2(x, y, z),\quad T_1z = \psi_3(x, y, z),$$

where ψ_1, ψ_2, ψ_3 are defined for every point of space and are single-valued.

34. Consider* the one-one point correspondence (T) defined by

$$Tx = \phi_1 (x, y, z, a_1, a_2, \ldots a_r), \quad Ty = \phi_2 (x, y, z, a_1, a_2, \ldots a_r),$$
$$Tz = \phi_3 (x, y, z, a_1, a_2, \ldots a_r),$$

where $a_1, a_2, \ldots a_r$ are r parameters. Let the parameters be assumed to be effective, so that two different choices of special values for them necessarily produce different correspondences. Then by varying the parameters an assemblage of correspondences is produced, each correspondence being defined by a particular choice of the parameters

$$a_1, a_2, \ldots a_r.$$

Now let S and T be any two members of this assemblage. Then STx, *i.e.* $S(Tx)$, STy, and STz obviously are the coordinates of a point which is related to the point (x, y, z) by a one-one point correspondence. This correspondence is denoted by ST. Now, if ST is necessarily a member of the assemblage whenever S and T are both members of it, the assemblage is called a group. When each correspondence of the group is defined by r effective parameters, where r is a finite number, the group is called finite and r-limbed. The group is said to be continuous, if, S and T being any two different transformations of the group, whenever the parameters of S vary continuously and ultimately approach those of T as their limits, then, for every value of x, y, z, also Sx, Sy, Sz vary continuously and approach Tx, Ty, Tz as their limits.

The assumption that ϕ_1, ϕ_2, ϕ_3 are analytical functions of their arguments, x, y, z, $a_1, \ldots a_r$, secures that the group is continuous.

35. The identical one-one point correspondence, Ω say, is such hat, for every value of x, y, z,

$$\Omega x = x, \quad \Omega y = y, \quad \Omega z = z \qquad \text{.......................(1)}.$$

Finite Continuous Transformation Groups exist which do not contain the identical transformation. But the chief interest of the ubject is concerned with those which do contain it. Let $a_1^0, a_2^0, \ldots a_r^0$ be the value-system of the parameters for which the corresponding ransformation of the group is the identical transformation Ω, so that

$$\Omega x = x = \phi_1 (x, y, z, a_1^0, a_2^0, \ldots a_r^0) \qquad \text{..............(2)},$$

with two similar equations.

* Cf. *Vorlesungen über Continuierliche Gruppen*, by Lie, ch. VI. § 2.

For brevity put

$$\left.\begin{aligned}\left(\frac{\partial\phi_1(x, y, z, a_1, \dots a_r)}{\partial a_n}\right)_{a_1^0, \dots a_r^0} &= \xi_n(x, y, z) = \xi_n, \ (n = 1, 2, \dots r)\\ \left(\frac{\partial\phi_2(x, y, z, a_1, \dots a_r)}{\partial a_n}\right)_{a_1^0, \dots a_r^0} &= \eta_n(x, y, z) = \eta_n, \ (n = 1, 2, \dots r)\\ \left(\frac{\partial\phi_3(x, y, z, a_1, \dots a_r)}{\partial a_n}\right)_{a_1^0, \dots a_r^0} &= \zeta_n(x, y, z) = \zeta_n, \ (n = 1, 2, \dots r)\end{aligned}\right\} \quad (3).$$

Now any transformation (T) of the group can be expressed in the form

$$\left.\begin{aligned}Tx &= \phi_1(x, y, z, a_1^0 + e_1 t, a_2^0 + e_2 t, \dots, a_r^0 + e_r t)\\ Ty &= \phi_2(x, y, z, a_1^0 + e_1 t, a_2^0 + e_2 t, \dots, a_r^0 + e_r t)\\ Tz &= \phi_3(x, y, z, a_1^0 + e_1 t, a_2^0 + e_2 t, \dots, a_r^0 + e_r t)\end{aligned}\right\} \quad \dots\dots\dots(4).$$

Hence, since the functions ϕ_1, ϕ_2, ϕ_3 are analytic, if t is not too large, we find, remembering equations (2) and (3),

$$\left.\begin{aligned}Tx &= x + t(e_1\xi_1 + e_2\xi_2 + \dots + e_r\xi_r) + \text{terms involving } t^2, t^3, \text{etc.}\\ Ty &= y + t(e_1\eta_1 + e_2\eta_2 + \dots + e_r\eta_r) + \text{terms involving } t^2, t^3, \text{etc.}\\ Tz &= z + t(e_1\zeta_1 + e_2\zeta_2 + \dots + e_r\zeta_r) + \text{terms involving } t^2, t^3, \text{etc.}\end{aligned}\right\} \quad (5).$$

Hence in the limit when t diminishes indefinitely, writing

$$Tx = x + \frac{dx}{dt} t, \text{ etc.},$$

we find

$$\left.\begin{aligned}\frac{dx}{dt} &= e_1\xi_1 + e_2\xi_2 + \dots + e_r\xi_r\\ \frac{dy}{dt} &= e_1\eta_1 + e_2\eta_2 + \dots + e_r\eta_r\\ \frac{dz}{dt} &= e_1\zeta_1 + e_2\zeta_2 + \dots + e_r\zeta_r\end{aligned}\right\} \dots\dots\dots\dots\dots\dots\dots(6).$$

These equations define the infinitesimal transformations of the group, every value-system of ratios of $e_1, e_2, \dots e_r$ defining one Infinitesimal Transformation.

36. Conversely by integrating equations (6) of § 35, it can be proved that the form of any finite transformation of the group can be recovered. Assume that we have found in this way

$$x = f_1(t, C_1, C_2, C_3), \quad y = f_2(t, C_1, C_2, C_3), \quad z = f_3(t, C_1, C_2, C_3),$$

where C_1, C_2, C_3 are the constants introduced by the integration. Let

x_0, y_0, z_0 be the values of x, y, z when $t=0$. Then from the above equations it can be proved to be possible, owing to the properties of the continuous group of one-one transformations, to solve for C_1, C_2, C_3 in terms of x_0, y_0, z_0. Thus we obtain again equations (4) of § 35, namely

$$x = \phi_1 (x_0, y_0, z_0, a_1^0 + e_1 t, \dots a_r^0 + e_r t),$$

with two similar equations, where x_0, y_0, z_0 now correspond to x, y, z in those equations, and x, y, z to Tx, Ty, Tz.

But it is not true that, if any equations of the same form as equations (6) of § 35 be written down, where $\xi_1, \dots, \zeta_r$ are any arbitrarily chosen functions of x, y, z, the integral forms give the finite transformations of an r-limbed finite continuous group. For in equations (6) of § 35, $\xi_1, \dots, \zeta_r$ are derived from equations (3) of § 35, that is to say, they are partial differential coefficients of functions with special properties. The enunciation of the conditions to be satisfied by $\xi_1, \dots, \zeta_r$, so that a *finite continuous* group of transformations may result from the integration of the corresponding equations, is called the Second Fundamental Theorem of the subject. It is not required here.

Also if $e_1, e_2, \dots e_r$ are kept unchanged, then the assemblage of transformations found by the variation of t in equations (4) of § 35 form a one-limbed continuous group, which is defined by the single infinitesimal transformation which it contains, namely that one corresponding to the given value-system of $e_1, e_2, \dots e_r$. Also every finite transformation is a member of the one-limbed group produced by the indefinite repetition of some infinitesimal transformation.

37. A latent point of a transformation is a point which is transformed into itself. A latent curve or surface is such that any point of it is transformed into some point on the same curve or surface.

It is evident that the latent points, lines, and surfaces of any infinitesimal transformation are also latent for every finite transformation belonging to the one-limbed group defined by it. They are called the latent points, lines, and surfaces of the group.

The transformations which leave a given surface latent must form a group, for the successive application of two such transformations still leaves the surface latent. Also the assemblage of the infinitesimal transformations which leave a surface latent must be the infinitesimal transformations of a continuous transformation group.

38. If a reentrant single-branched curve α (which may be a straight line) is transformed by an infinitesimal transformation of a continuous group into a curve β, then the senses* round, or directions round, the curves correspond in a perfectly definite manner, the same for all such infinitesimal transformations.

In order to make clear the correspondence of directions round any two reentrant single-branched curves α and β, let OP_1L and OP_2L define two complementary segments round α, and let $O'Q_1M$ and $O'Q_2M$ define two complementary segments round β. Now consider any one-one point transformation which (1) transforms α into β, (2) transforms segments of α into segments of β, (3) transforms O into O'. Then one of the two following mutually exclusive cases must hold, *either* one of the two, $O'Q_1M$ and the relatum of OP_1L, contains the other, *or* one of the two, $O'Q_2M$ and the relatum of OP_1L contains the other. If one of the two, $O'Q_1M$ and the relatum of OP_1L, contains the other, then the segments OP_1L and $O'Q_1M$ will be said to correspond in sense where O and O' are corresponding origins. Also we shall consider an arbitrary small portion of α containing O as the *neighbourhood* of O; thus O divides its neighbourhood into two parts, one lying in the segment OP_1L, and the other in the segment OP_2L. Similarly O' divides its neighbourhood on β into two parts. Then the case contemplated above, when the segments OP_1L and $O'Q_1M$ correspond in sense with O and O' as corresponding origins, will also be expressed by saying that O corresponds to O' and the neighbourhood of O in the segment OP_1L corresponds to the neighbourhood of O' in the segment $O'Q_1M$.

Now considering the case of an infinitesimal transformation, the curve β must lie infinitesimally near to the curve α, so that the point Q_1 may be assumed to be a point infinitesimally near to the point P_1 and the point Q_2 to be a point infinitesimally near to the point P_2. Then no point of the segment OP_1L which is infinitesimally near to P_1 is infinitesimally near to any point on the segment $O'Q_2M$. Hence the segments OP_1L and $O'Q_1M$ must correspond in sense with O and O' as corresponding origins. Thus only one of the two cases of correspondence in sense is now possible.

Notice that for this theorem the curves α and β need not be distinct, nor need the points O and O'.

If a straight line l is latent for a transformation, and O is a latent point on it, and segments with origin O correspond in sense with

* Cf. Proj. Geom. § 15, extended to any reentrant lines.

themselves, then the line is said to be transformed directly in the neighbourhood of O, in the other case it is said to be transformed inversely in the neighbourhood of O.

Thus it follows as a corollary from the above proposition that an infinitesimal transformation, which leaves latent a line and also a point O on it, transforms the line directly in the neighbourhood of O. Hence also it follows that any finite transformation of the one-limbed group defined by the infinitesimal transformation, transforms the line directly in the neighbourhood of O.

Similar theorems hold with respect to surfaces. It is sufficient for us to consider a transformation for which (1) a given straight line l is latent and also a point O on it, and (2) the relata of planes through l are planes and the relata of straight lines through O are straight lines. The general extension is obvious.

The portion of a plane through O, which lies within an arbitrarily small convex surface (cf. § 11) which contains O within it, will be

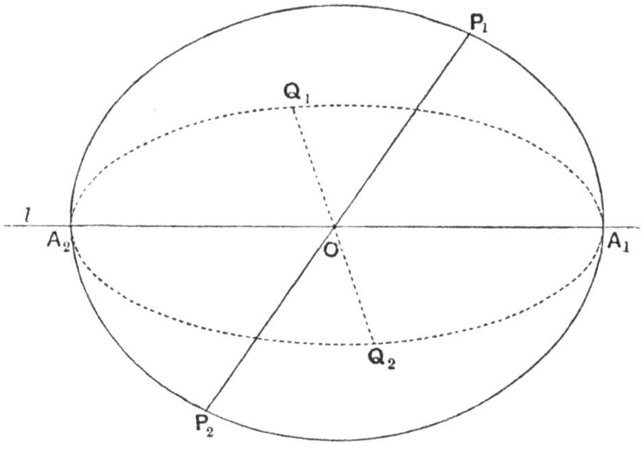

called the neighbourhood of O. The axis l divides into two parts the neighbourhood of O on a plane p through l; call them p_1 and p_2. Let the plane q be the relatum of p with respect to the transformation, and let the two parts of its neighbourhood, as divided by l, be q_1 and q_2. Let a line through O in p cut the convex surface in P_1, P_2; and let the relatum of the line in q cut the surface in Q_1, Q_2; also let OP_1 stand for the segment of the line in the neighbourhood p_1, and so on.

Then (assuming that continuous lines are transformed into continuous lines) if OP_1 and OQ_1 correspond in sense, the same must hold for all similar parts of corresponding lines through O in the neighbourhoods p_1 and q_1. The neighbourhoods p_1 and q_1 will then be said to correspond in sense. Also if p is latent, it will be said to be transformed directly in the neighbourhood of O with l as axis, if the neighbourhood p_1 corresponds to itself in sense.

Now, if the transformation is infinitesimal, it follows at once from the case of curves, that a definite one of the two neighbourhoods q_1 and q_2 must correspond in sense with p_1, and that, if the plane p is latent, it must be transformed directly in the neighbourhood of O with l as axis.

39. The general projective group of one-one point correspondences is the group of those transformations which transform planes into planes. Such transformations must therefore transform straight lines into straight lines, and must leave unaltered all projective relations between sets of points on lines.

Now, if in such a transformation three points A, B, C on a line l are known to be transformed into A', B', C' on a line l', the relatum on l' of every point on l is determined. For, by the Fundamental Theorem* one and only one projective relation can be established between the points on l and those on l', such that A corresponds to A', B to B', and C to C'. Thus the given transformation must transform l into l' according to this relation.

Hence it follows that if four points, A, B, C, D on a plane p, no three of which are collinear, are known to be transformed into A', B', C', D' on a plane p', the relatum on p' of every point on p is determined. For let AD meet BC in E, and $A'D'$ meet $B'C'$ in E'. Then E' corresponds to E. Hence A, B, E on AB correspond to A', B', E' on $A'B'$. Hence the relatum on $A'B'$ of every point on AB is determined, and similarly for BC and $B'C'$, and for CA and $C'A'$. But through any point P on p a line l can be drawn cutting BC, CA, AB in L, M, N. Thus the relata on p' of L, M, N, namely L', M', N' on l', are determined. Thus the relatum of every point on l is determined. Hence the relatum of P is determined.

Similarly, if A, B, C, D, E are five points, no four of which are coplanar, and if for any projective transformation their relata are determined, then the relatum of every point is determined. Accordingly a projective transformation is completely determined when the relata of five points, no four of which are coplanar, are determined.

* Cf. Proj. Geom. § 9 (γ).

40. Now consider transformations of the type

$$\left.\begin{aligned} Tx &= (a_{11}x + a_{12}y + a_{13}z + a_{14})/(a_1x + a_2y + a_3z + 1) \\ Ty &= (a_{21}x + a_{22}y + a_{23}z + a_{24})/(a_1x + a_2y + a_3z + 1) \\ Tz &= (a_{31}x + a_{32}y + a_{33}z + a_{34})/(a_1x + a_2y + a_3z + 1) \end{aligned}\right\} \quad \ldots(1).$$

They obviously belong to the general projective group as defined above. Also there are fifteen effective parameters. But if we substitute for x, y, z the coordinates of a given point A, and for Tx, Ty, Tz the coordinates of a given point A', three equations are obtained between the parameters. Let the same be done for B and B', C and C', D and D', E and E'. Then in all fifteen equations are found. Also if no four of A, B, C, D, E are coplanar, and no four of A', B', C', D', E' are coplanar, these equations are consistent, and definitely determine the transformation T. Hence (cf. § 39) the equations (1) can, by a proper choice of parameters, be made to represent any assigned transformation of the general projective group. Hence the transformations represented by them are those of the whole general projective group.

It is obvious from the form of these equations that the group is a fifteen-limbed continuous transformation-group. To find its infinitesimal transformations, put

$$a_{11} = 1 + \alpha_{11}t,\ a_{12} = \alpha_{12}t,\ a_{13} = \alpha_{13}t,\ a_{14} = \alpha_{14}t,\ a_1 = \alpha_1 t,\ a_2 = \alpha_2 t,\ a_3 = \alpha_3 t, \text{ etc.}$$

Then we find that the analogues of equations (6) of § 35 are

$$\left.\begin{aligned} \frac{dx}{dt} &= \alpha_{11}x + \alpha_{12}y + \alpha_{13}z + \alpha_{14} - x(\alpha_1x + \alpha_2y + \alpha_3z) \\ \frac{dy}{dt} &= \alpha_{21}x + \alpha_{22}y + \alpha_{23}z + \alpha_{24} - y(\alpha_1x + \alpha_2y + \alpha_3z) \\ \frac{dz}{dt} &= \alpha_{31}x + \alpha_{32}y + \alpha_{33}z + \alpha_{34} - z(\alpha_1x + \alpha_2y + \alpha_3z) \end{aligned}\right\} \quad \ldots\ldots(2).$$

These equations give the general form of an infinitesimal transformation of the general projective group.

CHAPTER V.

AXIOMS OF CONGRUENCE.

41. THE logical analysis of the method of superposition as applied to geometrical proofs is now to be undertaken. In this method a figure is said to move unchanged till it arrives at coincidence with some other figure. But what moves? Certainly not the points of the space. For they remain where they are. If it is some physical body occupying space which moves, then the assumption, that the body remains *unchanged* in its motion, involves the very comparison between the assemblage of points occupied in one position with that occupied in another position, which the supposition was designed to explain. Accordingly we find that Pasch* in effect treats 'congruence' as a fundamental idea not definable in terms of the geometrical concepts which we have already acquired. He states ten axioms of congruence in a form applicable to Descriptive Geometry. They are as follows, where the single capital letters represent points, and the figures are the ordered assemblages of the points mentioned, ordered in the order of mention.

I. The figures AB and BA are congruent.

II. To the figure ABC, one and only one point B' can be added, so that AB and AB' are congruent figures and B' lies in the segment AC or C in the segment AB'.

III. If the point C lies in the segment AB and the figures ABC and $A'B'C'$ are congruent, then the point C' lies in the segment $A'B'$.

IV. If the point C_1 lies in the segment AB, and the segment AC_1 is lengthened by the segment C_1C_2 which is congruent to it, and AC_2 is lengthened by the segment C_2C_3, congruent to AC_1, and so on, then finally a segment C_nC_{n+1} is arrived at which contains the point B.

* *loc. cit.* § 13.

V. If in the figure ABC the segments AC and BC are congruent, then the figures ABC and BAC are congruent.

VI. If two figures are congruent, so also are their homologous parts congruent.

VII. If two figures are each congruent to a third figure, they are congruent to each other.

VIII. If of two congruent figures one is enlarged by the addition of a point, the other can be similarly enlarged so that the enlarged figures are congruent.

IX. If AB and FGH are any two given figures, F, G, H being not collinear, and AB is congruent to FG, then in any plane containing AB exactly two points C and D can be found such that the figures ABC and ABD are each congruent to FGH, and furthermore the segment CD has a point in common with the line AB.

X. Two figures $ABCD$ and $ABCE$ which are not plane figures are not congruent.

42. These axioms at once suggest the analysis and definition of congruence in terms of our previously stated geometrical concepts. This analysis was first successfully achieved by Lie*.

Any point of space may be supposed to move with the rigid figure when the method of superposition is applied. Accordingly, considering the explanations of chapter IV, we see at once that a superposition is in fact a one-one point transformation. Let this special class of point transformations be called motions. We have now to consider whether the peculiar properties of motions can be defined in terms of the geometrical ideas already on hand.

If a rigid body is transferred from position α to position β, and then from β to γ, the final transformation defined is the same as if it were transferred directly from α to γ. Thus the successive application of two motions produces a motion. But this is the characteristic group property.

What Lie has succeeded in doing is to define in geometrical terms the properties which must be possessed by a complete group of motions. But now the explanations of the preceding paragraphs are found to be

* Cf. two papers by Lie in the Leipziger Berichte, 1890. These investigations are reproduced in a much enlarged form in the *Theorie der Transformationsgruppen*, vol. III. part V. But Lie's line of thought was not that suggested above. He starts from an almost successful solution of the same problem by Helmholtz, cf. *Ueber die Thatsachen, die der Geometrie zu Grunde liegen*, Gött. Nachr. 1868, and Collected Works, vol. II.

to some extent faulty. For they implicitly assume that there is one definite group of motions, as indeed our sensations of the physical world do in fact seem to give us special intelligence of one such definite group in physical space. However it will be found that an indefinite number of groups of one-one point transformations exist which satisfy Lie's definitions of the properties of a complete group of motions. Accordingly a motion when one special group is being considered is not a motion when another such group is considered.

A group of motions is called a congruence-group, and the definitions of the characteristics of such a group are called the axioms of Congruence.

43. Lie's results, as expressed by himself, are as follows:

*Definition**. A finite continuous group in the variables $x_1, x_2, \ldots x_n$ is called transitive, if in the space $(x_1, x_2, \ldots x_n)$ an n-fold extended region exists, within which each point can be transformed into any other point through at least one transformation of the group.

Definition†. A real continuous group of three-fold extended space possesses at the real point P free mobility in the infinitesimal, if it satisfies the following conditions: If a point P and an arbitrary real line-element passing through it are fixed, continuous motion is still possible; but if, in addition to P and that line-element, an arbitrary real surface-element, passing through both is held fixed, then shall no continuous motion be further possible.

Theorem‡. (1) If a real continuous projective group of ordinary three-fold extended space possesses without exception in all real points of this space free mobility in the infinitesimal, then it is six-limbed and transitive, and consists of all real projective transformations through which a not-exceptional imaginary surface of the second degree, which is represented by a real equation [*e.g.* $x^2 + y^2 + z^2 + 1 = 0$], remains invariant (latent).

(2) If a real continuous projective group of ordinary three-fold extended space possesses free mobility in the infinitesimal, not in all real points of this space but only in all real points of a certain region, then it is six-limbed and transitive and is either the continuous real projective group of a not-exceptional real not-ruled surface of the

* Cf. *Theorie der Transformationsgruppen*, vol. I. § 58.

† Cf. *loc. cit.*, vol. III. § 98.

‡ Cf. Lie, *loc. cit.* vol. III. § 98.

second degree [*i.e.* with this surface latent], or it is by means of a real projective transformation similar to the group of Euclidean motions.

The above constitutes what Lie calls his 'first solution of the Riemann*-Helmholtz Problem.'

The axioms which are implicit in this solution appear to be the following:

(1) A congruence-group is a finite continuous group of one-one point transformations, containing the identical transformation.

(2) It is a sub-group of the general projective group.

(3) An infinitesimal transformation belonging to it can always be found satisfying the condition, that any definite line and any definite point on the line are latent.

(4) No infinitesimal transformation of the group exists such that a line, a point on it, and a plane through it, shall all be latent.

44. Lie's† 'second solution of the Riemann-Helmholtz Problem' consists of the theorem that the following axioms completely characterize a complete assemblage of Euclidean or non-Euclidean Motions:

(1) The motions form a real continuous group defined by infinitesimal transformations.

(2) If any arbitrary real point (y_1^0, y_2^0, y_3^0) is fixed, then the real points (x_1, x_2, x_3), into which it is possible to move any real point (x_1^0, x_2^0, x_3^0), satisfy a real equation of the form

$$W(y_1^0, y_2^0, y_3^0;\ x_1^0, x_2^0, x_3^0;\ x_1, x_2, x_3) = 0,$$

which is not satisfied by $x_1 = y_1^0$, $x_2 = y_2^0$, $x_3 = y_3^0$, and which represents a real surface passing through (x_1^0, x_2^0, x_3^0).

(3) Round any point (y_1^0, y_2^0, y_3^0) a finite three-fold region exists, such that, when (y_1^0, y_2^0, y_3^0) is fixed, any other point (x_1^0, x_2^0, x_3^0) can be moved through an irreducible continuous sequence of points up to any point satisfying the above equation of (2).

45. The conception of a *finite* continuous group, though it is simple enough analytically, does not seem to correspond to any of the obvious and immediate properties of congruence-transformations as presented by sense-perceptions. The following set of axioms conform more closely to the obvious properties of congruence-transformations;

* Riemann's work in this connection is contained in his Habilitationsrede, *Ueber die Hypothesen, welche der Geometrie zu Grunde liegen*, 1854, cf. his Collected Works, and also a translation in the Collected Works of W. K. Clifford.

† Cf. *loc. cit.* vol. III. § 102.

they are based upon, and are modifications of, a set of congruence-axioms given by Peano*.

(1) The assemblage of congruence-transformations is a sub-group of the general projective group.

(2) The group contains the converse of every transformation belonging to it.

(3) Given any two points O and O', and any two lines l and l' through O and O' respectively, and any two planes π and π' through l and l' respectively, one and only one transformation of the group exists which transforms O into O', l into l', π into π', so that the two neighbourhoods of O on l correspond in an assigned manner with the two neighbourhoods of O' on l', and the two neighbourhoods of O on π as divided by l correspond in an assigned manner with the two neighbourhoods of O' on π' as divided by l'.

(4) Given any line and any point on that line, an infinitesimal transformation of the group exists such that the line and the point are latent.

Comparing these axioms with those of § 43 which are required for Lie's 'first solution,' it will be found that practically 'finite and continuous' is left out of the first axiom of § 43, but on the other hand the fourth axiom is strengthened into the form of axiom (3) of this article.

The following chapters will be based upon these axioms.

Proposition. It follows immediately from axioms (2) and (3) that the identical transformation is the only member of the group for which a given point is latent, and a given line through the point is latent, being transformed directly in the neighbourhood of the point, and a given plane through the line is latent, being transformed directly in the neighbourhood of the point with respect to the line as axis.

For with the notation of axiom (3) let T be such a transformation with respect to the point O, the line l, and the plane π. Also let S be the transformation of the group which transforms O, l, and π, into O', l', and π', in a specified way according to axiom (3); and let S_1 be the converse of S which also belongs to the group. Then the transformation ST belongs to the group, and transforms O, l, π into O', l', π' according to the same specified way as S. Hence by axiom (3), we have $ST = S$. Thus operating with S_1, we have $S_1ST = S_1S$. But by axiom (2) S_1ST and S_1S belong to the group; also $S_1ST = \Omega T = T$, and $S_1S = \Omega$. Hence $T = \Omega$.

* Cf. *loc. cit.* Riv. Mat. vol. IV.

CHAPTER VI.

INFINITESIMAL ROTATIONS.

46. An infinitesimal transformation of the projective group (cf. § 40, equations (2)), which leaves the origin and the axis of x latent, is of the form

$$\left.\begin{aligned} \frac{dx}{dt} &= a_{11}x + a_{12}y + a_{13}z - x(a_1x + a_2y + a_3z) \\ \frac{dy}{dt} &= a_{22}y + a_{23}z - y(a_1x + a_2y + a_3z) \\ \frac{dz}{dt} &= a_{32}y + a_{33}z - z(a_1x + a_2y + a_3z) \end{aligned}\right\} \quad \text{.........(1).}$$

We proceed to consider the specialization necessary for the coefficients in order that this may be a 'rotation round the axis of x' in a congruence group.

There is in a congruence group only one infinitesimal 'rotation' round any given line with a given point on the line latent. For consider the motion of the plane, $y = pz$, round the axis of x with the origin latent; after the infinitesimal transformation (1), we have

$$\frac{dy}{dt} = p\frac{dz}{dt} + z\frac{dp}{dt}.$$

Substituting from (1), and putting $y = pz$, we find

$$\frac{dp}{dt} = a_{23} + (a_{22} - a_{38})\,p - a_{32}p^2.$$

Hence when p is changed to $p + dp$ by the infinitesimal transformation (1) we find

$$d_1t = \frac{dp}{a_{23} + (a_{22} - a_{33})\,p - a_{32}p^2} \quad \text{.....................(2).}$$

Now consider a second infinitesimal transformation of similar form to (1), only with a_{11}', a_{12}', etc. as coefficients. Let d_2t be the increment of t requisite to change p into $p+dp$. Then we have

$$d_2t = \frac{dp}{a_{23}' + (a_{22}' - a_{33}')p - a_{32}'p^2} \quad \text{.................(3).}$$

Now consider the transformation

$$dx = \left(\frac{dx}{dt}\right)_1 d_1t - \left(\frac{dx}{dt}\right)_2 d_2t,$$

with two similar equations; where $\left(\frac{dx}{dt}\right)_1$ comes from the first transformation, and $\left(\frac{dx}{dt}\right)_2$ from the second. But this transformation leaves the plane, $y=pz$, latent. Hence by the proposition of § 45, it is the identical transformation. Thus we find $dx=0$, $dy=0$, $dz=0$, for every value of x, y, z, and p. Thus

$$\frac{a_{11}x + a_{12}y + a_{13}z - x(a_1x + a_2y + a_3z)}{a_{23} + (a_{22} - a_{33})p - a_{32}p^2} = \frac{a_{11}'x + a_{12}'y + a_{13}'z - x(a_1'x + a_2'y + a_3'z)}{a_{23}' + (a_{22}' - a_{33}')p - a_{32}'p^2},$$

with corresponding equations for y and z. These three equations hold for every value of x, y, z, and p. Hence it is easy to prove that

$$\frac{a_{11}'}{a_{11}} = \frac{a_{12}'}{a_{12}} = \ldots = \frac{a_{33}'}{a_{33}} = \ldots = \frac{a_3'}{a_3} \quad \text{.................(4).}$$

Thus the infinitesimal transformations are identical.

47. The plane, $my+nz=0$, is latent for the rotation of § 46 (1), if, $m\frac{dy}{dt} + n\frac{dz}{dt} = 0$, is satisfied whenever the point (x, y, z) lies on the plane. Hence

$$ma_{22} + na_{32} = \sigma m,$$
$$ma_{23} + na_{33} = \sigma n,$$

and σ is given by

$$(\sigma - a_{22})(\sigma - a_{33}) - a_{23}a_{32} = 0 \quad \text{...................(1).}$$

But by the proposition of § 45, there can be no real latent plane of this form. Hence the roots of equation (1) are imaginary. Thus

$$4(a_{22}a_{33} - a_{23}a_{32}) - (a_{22} + a_{33})^2 > 0 \quad \text{.................(2).}$$

48. In the neighbourhood of the origin the rotations of § 46 (1) can be expressed by

$$\left.\begin{aligned}\frac{dx}{dt} &= a_{11}x + a_{12}y + a_{13}z\\ \frac{dy}{dt} &= \qquad\quad a_{22}y + a_{23}z\\ \frac{dz}{dt} &= \qquad\quad a_{32}y + a_{33}z\end{aligned}\right\} \quad \text{.....................(1).}$$

Thus, writing $y \propto e^{pt}$, $z \propto e^{pt}$, p satisfies equation (1) of § 47. Hence p is complex. Thus we may write

$$\left.\begin{aligned} y &= e^{\frac{1}{2}(a_{22}+a_{33})t}\,(y_0 \cos \nu t + q \sin \nu t)\\ z &= e^{\frac{1}{2}(a_{22}+a_{33})t}\,(z_0 \cos \nu t + q' \sin \nu t)\end{aligned}\right\} \quad \text{...............(2),}$$

where $\nu = \sqrt{\{a_{22}a_{33} - a_{23}a_{32} - \frac{1}{4}(a_{22} + a_{33})^2\}}$, and q and q' can be determined in terms of y_0, z_0, and of the coefficients. Thus, putting $\lambda = e^{\frac{1}{2}(a_{22}+a_{33})\pi/\nu}$, when $t = \pi/\nu$,

$$y = -\lambda y_0, \quad z = -\lambda z_0 \quad \text{.......................(3),}$$

and when $t = 2\pi/\nu$,

$$y = \lambda^2 y_0, \quad z = \lambda^2 z_0 \quad \text{.......................(4).}$$

By the proposition of § 45, the equations (4) must reduce to $y = y_0$, $z = z_0$. Hence $\lambda = 1$, and therefore

$$a_{22} + a_{33} = 0 \quad \text{..........................(5).}$$

Thus for a value of t, not zero, the integral form of equations (1) yields the identical transformation.

Also equations (2) become

$$\left.\begin{aligned} y &= y_0 \cos \nu t + \frac{a_{22}y_0 + a_{23}z_0}{\nu} \sin \nu t\\ z &= z_0 \cos \nu t + \frac{a_{32}y_0 + a_{33}z_0}{\nu} \sin \nu t\end{aligned}\right\} \quad \text{...............(6).}$$

Hence a value of t can be found such that by the corresponding transformation of the type of equations (6), any plane $y_0 = p_0 z_0$ is transformed into any plane $y = pz$, the axis of x being transformed directly, and the neighbourhoods of O on the planes as divided by the axis of x corresponding in assigned manners. Hence by axiom (3) of § 45, this is the only transformation of the group for which these conditions are fulfilled.

Hence the transformations for which the origin and the axis of x are latent, the axis of x being transformed directly in the neighbourhood of the origin, form a one-limbed continuous group produced by the infinitesimal transformation which fulfils these conditions.

49. The transformation of § 46 (1) on the latent axis of x (*i.e.* $y=0$, $z=0$) is given by

$$\frac{dx}{dt}=a_{11}x-a_1x^2.$$

If $a_{11}\neq 0$, the solution is

$$\frac{x}{a_{11}-a_1x}=\frac{x_0}{a_{11}-a_1x_0}e^{a_{11}t}.$$

If $a_{11}=0$, the solution is

$$\frac{1}{x}-\frac{1}{x_0}=a_1 t.$$

But (cf. § 48) when $t=2\pi/\nu$, we find $x=x_0$ for every value of x_0. Hence

$$a_{11}=0, \quad a_1=0.$$

Thus every point on any line is latent for a rotation round it with one point of it latent. This fundamental theorem will be cited by the shortened statement, that 'every point on an axis of rotation is latent.'

Thus equations (1) of § 46 for the infinitesimal rotation round the axis of x, reduce to

$$\left.\begin{aligned}\frac{dx}{dt}&=a_{12}y+a_{13}z-x(a_2y+a_3z)\\ \frac{dy}{dt}&=a_{22}y+a_{23}z-y(a_2y+a_3z)\\ \frac{dz}{dt}&=a_{32}y+a_{33}z-z(a_2y+a_3z)\end{aligned}\right\}\quad\ldots\ldots\ldots\ldots\ldots(1),$$

where and

$$\left.\begin{aligned}a_{22}\quad+a_{33}\quad&=0\\ a_{22}a_{33}-a_{23}a_{32}&>0\end{aligned}\right\}\quad\ldots\ldots\ldots\ldots\ldots\ldots\ldots\ldots(2).$$

50. The condition that

$$lx+my+nz=0,\ (l\neq 0)\quad\ldots\ldots\ldots\ldots\ldots\ldots\ldots(1),$$

should be a latent plane for the rotation (1) of § 49 is that

$$l\frac{dx}{dt}+m\frac{dy}{dt}+n\frac{dz}{dt}=0\ldots\ldots\ldots\ldots\ldots\ldots\ldots(2),$$

whenever (1) is satisfied. Hence substituting for

$$\frac{dx}{dt},\ \frac{dy}{dt},\ \frac{dz}{dt},$$

and using (1), we find

$$\left.\begin{aligned}a_{12}l+a_{22}m+a_{32}n&=0\\ a_{13}l+a_{23}m+a_{33}n&=0\end{aligned}\right\}\quad\ldots\ldots\ldots\ldots\ldots\ldots\ldots(3).$$

From the inequality (2) of § 49, it follows that the solution of this equation satisfies the condition, $l \neq 0$.

Let this plane be taken to be the plane of yz, *i.e.* the plane $x = 0$. This requires

$$\alpha_{12} = 0, \quad \alpha_{13} = 0 \quad \text{.........................(4).}$$

51. With this specialization of the plane of yz, the condition that,

$$lx + my + nz + 1 = 0 \quad \text{.......................(1),}$$

should be a latent plane for the rotation (1) of § 49 is that

$$l\frac{dx}{dt} + m\frac{dy}{dt} + n\frac{dz}{dt} = 0,$$

whenever (1) is satisfied. Hence substituting from equations (1) of § 49 and using equation (1), we find (cf. § 50, equation (4))

$$\left.\begin{aligned} \alpha_{22}m + \alpha_{32}n + \alpha_2 = 0 \\ \alpha_{32}m + \alpha_{33}n + \alpha_3 = 0 \end{aligned}\right\} \quad \text{.......................(2).}$$

Hence there is a family of latent planes of the form (1), where l is the variable parameter, and m and n are definitely determined in terms of the coefficients of the infinitesimal rotation. Now let one member of this family be taken to be the infinite plane. Then from equations (2), we find

$$\alpha_2 = 0, \quad \alpha_3 = 0 \quad \text{.........................(3).}$$

Hence with these choices for the plane of yz and for the infinite plane, the infinitesimal rotation round the axis of x is reduced to the form

$$\left.\begin{aligned} \frac{dx}{dt} &= 0 \\ \frac{dy}{dt} &= \alpha_{22}y + \alpha_{23}z \\ \frac{dz}{dt} &= \alpha_{32}y + \alpha_{33}z \end{aligned}\right\} \quad \text{.......................(4),}$$

where

$$\left.\begin{aligned} \alpha_{22} + \alpha_{33} &= 0 \\ \alpha_{22}\alpha_{33} - \alpha_{23}\alpha_{32} &> 0 \end{aligned}\right\} \quad \text{.........................(5).}$$

Then every plane of the family, $\lambda x + \mu = 0$, is latent.

52. Any infinitesimal motion, which keeps the origin fixed, is of the form

$$\left.\begin{aligned} &\frac{dx}{dt} = \epsilon_{11}x + \epsilon_{12}y + \epsilon_{13}z - x(\epsilon_1 x + \epsilon_2 y + \epsilon_3 z) \\ &\text{with two similar equations} \end{aligned}\right\} \quad \text{...........(1).}$$

If the line, $x = l\sigma$, $y = m\sigma$, $z = n\sigma$, is latent, then

$$\frac{dx}{dt} = l\frac{d\sigma}{dt}, \quad \frac{dy}{dt} = m\frac{d\sigma}{dt}, \quad \frac{dz}{dt} = n\frac{d\sigma}{dt}.$$

Hence putting ρ for $\frac{1}{\sigma}\frac{d\sigma}{dt}$, the equations (1) become

$$(\epsilon_{11} - \rho) l + \epsilon_{12} m + \epsilon_{13} n - \sigma l (\epsilon_1 l + \epsilon_2 m + \epsilon_3 n) = 0,$$

with two similar equations.

These equations hold for all values of σ. Accordingly, near the origin, when σ is very small,

$$(\epsilon_{11} - \rho) l + \epsilon_{12} m + \epsilon_{13} n = 0,$$

with two similar equations.

Hence ρ, in the neighbourhood of the origin, satisfies

$$\begin{vmatrix} \epsilon_{11} - \rho, & \epsilon_{12}, & \epsilon_{13} \\ \epsilon_{21}, & \epsilon_{22} - \rho, & \epsilon_{23} \\ \epsilon_{31}, & \epsilon_{32}, & \epsilon_{33} - \rho \end{vmatrix} = 0.$$

But this equation has always one real root. Thus there is always one real latent line through the origin. Hence every infinitesimal motion for which one point is latent possesses an 'axis.' Also (cf. § 49) every point on this axis is latent. Accordingly for every point on the axis, $x = l\sigma$, $y = m\sigma$, $z = n\sigma$, we have

$$\frac{dx}{dt} = 0, \quad \frac{dy}{dt} = 0, \quad \frac{dz}{dt} = 0.$$

Hence $$\epsilon_{11} l + \epsilon_{12} m + \epsilon_{13} n - \sigma (l\epsilon_1 + m\epsilon_2 + n\epsilon_3) = 0,$$

with two similar equations.

These equations hold for every value of σ. Thus

$$\left.\begin{aligned} \epsilon_{11} l + \epsilon_{12} m + \epsilon_{13} n &= 0 \\ \epsilon_{21} l + \epsilon_{22} m + \epsilon_{23} n &= 0 \\ \epsilon_{31} l + \epsilon_{32} m + \epsilon_{33} n &= 0 \\ \epsilon_1 l + \epsilon_2 m + \epsilon_3 n &= 0 \end{aligned}\right\} \quad \text{......................(2).}$$

Hence we find the equation, $|\epsilon_{rs}| = 0$, and that the values of $l : m : n$ which satisfy the first three equations, must satisfy the fourth.

53. The infinitesimal rotation round the axis of y as axis is of the form (cf. § 49, equations (1) and (2)),

$$\left.\begin{aligned}\frac{dx}{dt} &= \beta_{11}x + \beta_{13}z - x(\beta_1 x + \beta_3 z)\\ \frac{dy}{dt} &= \beta_{21}x + \beta_{23}z - y(\beta_1 x + \beta_3 z)\\ \frac{dz}{dt} &= \beta_{31}x + \beta_{33}z - z(\beta_1 x + \beta_3 z)\end{aligned}\right\} \quad \text{..............(1)},$$

where and

$$\left.\begin{aligned}\beta_{11} + \beta_{33} &= 0\\ \beta_{11}\beta_{33} - \beta_{13}\beta_{31} &> 0\end{aligned}\right\} \quad \text{......................(2)}.$$

Then, since (cf. § 45, axiom (1)) the motions form a group, by combining this infinitesimal rotation with that round the axis of x, another infinitesimal rotation of the group is found. Thus (cf. § 51, equations (4)) an infinitesimal rotation of the group, assuming the special axes and infinite plane of § 51, is of the form

$$\left.\begin{aligned}\frac{dx}{dt} &= \beta_{11}x + \beta_{13}z - x(\beta_1 x + \beta_3 z)\\ \frac{dy}{dt} &= \beta_{21}x + \kappa\alpha_{22}y + (\beta_{23} + \kappa\alpha_{23})z - y(\beta_1 x + \beta_3 z)\\ \frac{dz}{dt} &= \beta_{31}x + \kappa\alpha_{32}y + (\beta_{33} + \kappa\alpha_{33})z - z(\beta_1 x + \beta_3 z)\end{aligned}\right\} \quad \text{........(3)},$$

where κ has any arbitrary value.

Hence (cf. § 52) we have

$$\begin{vmatrix}\beta_{11}, & 0, & \beta_{13}\\ \beta_{21}, & \kappa\alpha_{22}, & \beta_{23} + \kappa\alpha_{23}\\ \beta_{31}, & \kappa\alpha_{32}, & \beta_{33} + \kappa\alpha_{33}\end{vmatrix} = 0 \quad \text{...............(4)}.$$

But equation (4) holds for every value of κ. Hence

$$\beta_{11}(\alpha_{22}\alpha_{33} - \alpha_{23}\alpha_{32}) = 0.$$

Hence (cf. § 51, equations (5)),

$$\beta_{11} = 0 \quad \text{..............................(5)}.$$

Thence, again from equation (4), we find

$$\beta_{13}\beta_{21}\alpha_{32} - \beta_{13}\beta_{31}\alpha_{22} = 0 \quad \text{......................(6)}.$$

From equations (2) and (5) we find

$$\beta_{33} = 0 \quad \text{..............................(7)}.$$

54. Now (cf. § 50) the plane of yz is the latent plane through the origin of the infinitesimal rotation round the axis of x, and the axes of y and z are any distinct lines in this plane through the origin. Any point on the latent plane, $lx + my + nz = 0$, of the rotation round the axis of y satisfies (cf. § 53, equations (1), (5), and (7))

$$l\beta_{13}z + m(\beta_{21}x + \beta_{23}z) + n\beta_{31}x = 0.$$

Hence $$m\beta_{21} + n\beta_{31} = 0, \quad l\beta_{13} + m\beta_{23} = 0.$$

Thus the equation of the latent plane is

$$\beta_{23}\beta_{31}x - \beta_{13}\beta_{31}y + \beta_{21}\beta_{13}z = 0 \quad \text{..................(1).}$$

But (cf. § 53, equations (2)) $\beta_{13}\beta_{31}$ cannot vanish. Hence the latent plane cannot contain the axis of y. Thus we may assume its intersection with the plane, $x = 0$ (*i.e.* with the latent plane of the rotation round Ox), to be the axis of z. With this assumption we have

$$\beta_{21} = 0 \quad \text{..................................(2).}$$

Then from equation (6) of § 53, we find

$$\alpha_{22} = 0 \quad \text{..................................(3).}$$

And from equations (5) of § 51, we find

$$\alpha_{33} = 0 \quad \text{..................................(4).}$$

A latent plane of an infinitesimal rotation round an axis will be said to be perpendicular to the axis. The set of axes of coordinates with any given origin, found by taking the axis of x to be any line, the axis of y to be any line in the latent plane through the origin of the infinitesimal rotation round the axis of x, and the axis of z to be the line of intersection of the latent planes through the origin of the infinitesimal rotations round the axes of x and y, will be said to be mutually perpendicular, or mutually at right angles.

It has now to be proved that a set of axes mutually at right angles have reciprocal properties in respect to each other.

55. With the mutually perpendicular axes of § 54, the equations (2) of § 52, as applied to the infinitesimal rotation of equations (3) of § 53, become

$$\beta_{13}n = 0, \quad (\beta_{23} + \kappa\alpha_{23})\,n = 0, \quad \beta_{31}l + \kappa\alpha_{32}m = 0,$$
$$\beta_1 l + \beta_3 n = 0.$$

Hence κ can be given any arbitrary value, and then the corresponding values of $l : m : n$ are to be found. Also β_{13}, β_{31}, α_{23}, α_{32} cannot vanish.

Hence we have $\quad n = 0, \quad l/\kappa\alpha_{32} = m/(-\beta_{31})$.

Thus
$$\beta_1 = 0 \quad \text{.................................(1).}$$

56. Again, let $lx + my + nz + 1 = 0$, be any one of the family of latent planes of the rotation round the axis of y. Then for all points on the plane,

$$l\frac{dx}{dt} + m\frac{dy}{dt} + n\frac{dz}{dt} = 0.$$

Hence substituting from equations (1) of § 53, remembering that

$$\beta_{11}, \quad \beta_{33}, \quad \beta_{21}, \quad \beta_1$$

have all been proved to vanish for the special axes, we have

$$n\beta_{31}x + (l\beta_{13} + m\beta_{23} + \beta_3)z = 0$$

for all points on the plane. Also β_{13}, β_{31} do not vanish. Hence

$$n = 0, \quad l = -(m\beta_{23} + \beta_3)/\beta_{13}.$$

Thus there is one latent plane for which $m = 0$, $n = 0$, $l = -\beta_3/\beta_{13}$. This is the plane

$$-\beta_3 x + \beta_{13} = 0.$$

But this plane is a member of the family (cf. § 51) of latent planes of the rotation round the axis of x. Also the infinite plane has been chosen to be any member of this family. Thus we now choose the infinite plane to be the one common member of the two families of latent planes of the infinitesimal rotations round the axes of x and of y. This plane, since $\beta_{13} \neq 0$, can never pass through the origin. With this choice, we find

$$\beta_3 = 0 \quad \text{.................................(1).}$$

Then, with this special tetrahedron of reference, the equations defining the infinitesimal rotation round the axis of x are reduced to

$$\frac{dx}{dt} = 0, \quad \frac{dy}{dt} = \alpha_{23}z, \quad \frac{dz}{dt} = \alpha_{32}y \quad \text{..................(2),}$$

where
$$\alpha_{23}\alpha_{32} < 0 \quad \text{..............................(3).}$$

Also the equations defining the infinitesimal rotation round the axis of y are reduced to

$$\frac{dx}{dt} = \beta_{13}x, \quad \frac{dy}{dt} = \beta_{23}z, \quad \frac{dz}{dt} = \beta_{31}x \quad \text{..............(4),}$$

where
$$\beta_{13}\beta_{31} < 0 \quad \text{..............................(5).}$$

57. The equations defining the infinitesimal rotation round the axis of z are (cf. § 53, equations (1), (5) and (7)),

$$\left.\begin{aligned}\frac{dx}{dt}&=\gamma_{12}y-x(\gamma_1x+\gamma_2y)\\ \frac{dy}{dt}&=\gamma_{21}x-y(\gamma_1x+\gamma_2y)\\ \frac{dz}{dt}&=\gamma_{31}x+\gamma_{32}y-z(\gamma_1x+\gamma_2y)\end{aligned}\right\}\quad\text{..............(1),}$$

where $$\gamma_{12}\gamma_{21}<0\quad\text{..............................(2).}$$

Thus the transformation found by combining three infinitesimal rotations round the axes of x, of y, and of z is by equations (2) and (4) of § 56,

$$\left.\begin{aligned}\frac{dx}{dt}&=\kappa_3\gamma_{12}y+\kappa_2\beta_{13}z-x(\kappa_3\gamma_1x+\kappa_3\gamma_2y)\\ \frac{dy}{dt}&=\kappa_3\gamma_{21}x+(\alpha_{23}+\kappa_2\beta_{23})z-y(\kappa_3\gamma_1x+\kappa_3\gamma_2y)\\ \frac{dz}{dt}&=(\kappa_2\beta_{31}+\kappa_3\gamma_{31})x+(\kappa_3\gamma_{32}+\alpha_{32})y-z(\kappa_3\gamma_1x+\kappa_3\gamma_2y)\end{aligned}\right\}\quad\text{......(3).}$$

Hence applying equations (2) of § 52, we find

$$\begin{vmatrix}0, & \kappa_3\gamma_{12}, & \kappa_2\beta_{13}\\ \kappa_3\gamma_{21}, & 0, & \alpha_{23}+\kappa_2\beta_{23}\\ \kappa_2\beta_{31}+\kappa_3\gamma_{31}, & \kappa_3\gamma_{32}+\alpha_{32}, & 0\end{vmatrix}=0.$$

This equation holds for every value of κ_2 and κ_3.

Thus the term involving $\kappa_2^2\kappa_3$ yields $\gamma_{12}\beta_{23}\beta_{31}=0$. Hence, since γ_{12} and β_{31} cannot vanish, we have

$$\beta_{23}=0\quad\text{..................................(4).}$$

The term involving κ_3^2 yields $\gamma_{12}\gamma_{31}\alpha_{23}=0$. Hence, since γ_{12} and α_{23} cannot vanish, we have

$$\gamma_{31}=0\quad\text{..................................(5).}$$

The coefficient of $\kappa_3^2\kappa_2$ is $\gamma_{12}\gamma_{31}\beta_{23}+\beta_{13}\gamma_{21}\gamma_{32}$. Hence, using (4) and (5), and noting that β_{13} and γ_{21} cannot vanish, we find

$$\gamma_{32}=0\quad\text{..................................(6).}$$

The coefficient of $\kappa_2\kappa_3$ gives

$$\gamma_{12}\alpha_{23}\beta_{31}+\beta_{13}\alpha_{32}\gamma_{21}=0\quad\text{........................(7).}$$

Now equations (2) of § 52, applied to this case, become, after simplifying by (4), (5) and (6),

$$\kappa_3\gamma_{12}m + \kappa_2\beta_{13}n = 0,$$
$$\kappa_3\gamma_{21}l + \alpha_{23}n = 0,$$
$$\kappa_2\beta_{31}l + \alpha_{32}m = 0,$$
$$\kappa_3\gamma_1 l + \kappa_3\gamma_2 m = 0.$$

Hence $$\alpha_{32}\gamma_1 - \kappa_2\beta_{31}\gamma_2 = 0.$$

This equation holds for all values of κ_2, and α_{32} and β_{31} do not vanish.

Hence $$\gamma_1 = 0, \quad \gamma_2 = 0 \quad \text{.........................(8).}$$

Thus the infinite plane is the common latent plane of the three infinitesimal rotations round the three rectangular axes.

58. Thus using equations (4), (5), (6), (8) of § 57, the equations for the infinitesimal rotations round the three mutually perpendicular axes, the infinite plane being the common latent plane of the rotations, are

$$\frac{dx}{dt} = 0, \quad \frac{dy}{dt} = \alpha_{23}z, \quad \frac{dz}{dt} = \alpha_{32}y \quad \text{.................(1),}$$

$$\frac{dx}{dt} = \beta_{13}z, \quad \frac{dy}{dt} = 0, \quad \frac{dz}{dt} = \beta_{31}x \quad \text{................(2),}$$

$$\frac{dx}{dt} = \gamma_{12}y, \quad \frac{dy}{dt} = \gamma_{21}x, \quad \frac{dz}{dt} = 0 \quad \text{.................(3),}$$

where $$\alpha_{23}\alpha_{32} < 0, \quad \beta_{13}\beta_{31} < 0, \quad \gamma_{12}\gamma_{21} < 0 \quad \text{.................(4),}$$

and $$\gamma_{12}\alpha_{23}\beta_{31} + \beta_{13}\alpha_{32}\gamma_{21} = 0 \quad \text{.......................(5).}$$

It at once follows from the symmetry of these equations, that a set of axes mutually at right angles have reciprocal properties in respect to each other.

The mention of equations (4) and (5) is avoided by altering the unit points* on the axes, that is, by writing λx for x, μy for y, and νz for z, where λ, μ, ν are constants at our disposal. Let them be chosen, so that

$$\mu\alpha_{32}/\nu = -\nu\alpha_{23}/\mu = \omega_1 \text{ (say),}$$

and $$\nu\beta_{13}/\lambda = -\lambda\beta_{31}/\nu = \omega_2 \text{ (say).}$$

By equations (4), the ratios of $\lambda : \mu : \nu$ are real. Then by equation (5), we have

$$\lambda\gamma_{21}/\mu = -\mu\gamma_{12}/\lambda = \omega_3 \text{ (say).}$$

* Cf. Proj. Geom. § 42.

Hence remembering that ω_1, ω_2, ω_3 are arbitrary parameters, we find that any infinitesimal rotation round an axis through the origin can be expressed in the form

$$\frac{dx}{dt} = -\omega_3 y + \omega_2 z, \quad \frac{dy}{dt} = -\omega_1 z + \omega_3 x, \quad \frac{dz}{dt} = -\omega_2 x + \omega_1 y \quad \ldots(6).$$

The latent line of the rotation is given by

$$x/\omega_1 = y/\omega_2 = z/\omega_3.$$

Thus this form gives one and only one infinitesimal rotation round any line through the origin. Hence the form (6) can include no infinitesimal transformation other than those of the congruence group under consideration.

A tetrahedron formed by three mutually perpendicular axes, with the common latent plane of the three rotations round the axes for its fourth plane, and with the unit points of its axes chosen so as to produce equations (6), will be called a normal reference tetrahedron.

When the congruence group is given, the normal reference tetrahedrons are determinate, though infinite in number. But a congruence group can be found so that any given tetrahedron is a normal reference tetrahedron.

CHAPTER VII.

THE ABSOLUTE.

59. CONSIDER the surfaces which are latent for a rotation round the axis of x. Let the axis system form a normal reference tetrahedron. Then the infinitesimal rotation can be written

$$\frac{dx}{dt}=0,\quad \frac{dy}{dt}=-\omega_1 z,\quad \frac{dz}{dt}=\omega_1 y \quad \text{..............(1)}.$$

Let $u=0$ be any latent surface. Then we have

$$-z\frac{\partial u}{\partial y}+y\frac{\partial u}{\partial z}=0,$$

as the requisite condition. Solving this linear equation by Lagrange's rule, and remembering that x has been treated as a constant, we find that the latent surfaces are of the form

$$f(y^2+z^2,\ x)=0 \text{ ....(2)},$$

where f is an arbitrary function. Surfaces, whose equations are of the form of (2), will be called surfaces of revolution round the axis of x.

60. A necessary and sufficient condition, that a surface may be latent for any congruence transformation which leaves the origin at rest, is that the surface be a surface of revolution round each of the three axes. Hence by equation (2) of § 59 this family of surfaces is represented by

$$\lambda(x^2+y^2+z^2)+\mu=0 \text{ ....(1)},$$

where λ and μ are arbitrary parameters. Let these be called spheres, with the origin as centre.

The infinite plane is the common polar plane of the origin, with respect to each of the spheres with it as centre. Thus transforming to homogeneous coordinates by putting $x = X/U$, $y = Y/U$, $z = Z/U$, where, $U = 0$, is the equation of the infinite plane, the equation of the family of concentric spheres is

$$\lambda (X^2 + Y^2 + Z^2) + \mu U^2 = 0 \quad \ldots\ldots\ldots\ldots\ldots\ldots(2).$$

Thus returning to the original coordinates, if $\phi (x, y, z) = 0$, is the equation of any sphere, centre at (x_0, y_0, z_0), the equation of the family of spheres with that centre is

$$\lambda\phi (x, y, z) + \mu \left(x_0 \frac{\partial \phi}{\partial x} + y_0 \frac{\partial \phi}{\partial y} + z_0 \frac{\partial \phi}{\partial z} + \frac{\partial \phi}{\partial t} \right)^2 = 0 \quad \ldots\ldots(3),$$

where, as usual, t is introduced to make the equation homogeneous and is put equal to 1 after differentiation.

61. By recurring to equation (2) of § 60, we see that the plane of yz, which is the plane perpendicular to the axis of x, is the plane through the origin and through the common conjugate line of the axis of x with respect to any of the spheres, centre the origin. Hence if, $\phi (x, y, z) = 0$, is any sphere with centre $A_0 (x_0, y_0, z_0)$, and A_1 is the point (x_1, y_1, z_1), then the plane through A_0 perpendicular to $A_0 A_1$ is

$$(x_0 + \lambda x_1) \frac{\partial \phi}{\partial x} + (y_0 + \lambda y_1) \frac{\partial \phi}{\partial y} + (z_0 + \lambda z_1) \frac{\partial \phi}{\partial z} + (1 + \lambda) \frac{\partial \phi}{\partial t} = 0 \quad \ldots(1),$$

where

$$\lambda = - 2\phi_0 \Big/ \left\{ x_1 \left(\frac{\partial \phi}{\partial x}\right)_0 + y_1 \left(\frac{\partial \phi}{\partial y}\right)_0 + z_1 \left(\frac{\partial \phi}{\partial z}\right)_0 + \left(\frac{\partial \phi}{\partial t}\right)_0 \right\},$$

and ϕ_0, $\left(\frac{\partial \phi}{\partial x}\right)_0$, etc. are the results of substituting the coordinates of A_0 in $\phi (x, y, z)$, $\frac{\partial \phi}{\partial x}$, etc.

Let the left-hand side of (1) be written $(A_0, A_1, P)_\phi$, where P is the variable point (x, y, z). Thus the equation of the plane, perpendicular to the line $A_0 A_1$ and through the point A_0, is

$$(A_0, A_1, P)_\phi = 0 \quad \ldots\ldots\ldots\ldots\ldots\ldots\ldots\ldots(2).$$

A quadratic surface of revolution round the axis of x is of the form (cf. equation (2) of § 59)

$$b (y^2 + z^2) + ax^2 + 2gx + c = 0 \quad \ldots\ldots\ldots\ldots\ldots\ldots(3).$$

This can be written in the form

$$\lambda \{\alpha (x^2 + y^2 + z^2) + \beta\} + a' x^2 + 2gx + c' = 0.$$

Thus, if $\phi (x, y, z) = 0$ is the equation of a sphere, centre $A_0 (x_0, y_0, z_0)$, the equation of a quadric surface of revolution round the line joining A_0 to A_1 (x_1, y_1, z_1) is

$$\lambda\phi(x, y, z) + \mu(A_0, A_1, P)^2_\phi + \nu(A_0, A_1, P)_\phi\left(x_0\frac{\partial\phi}{\partial x} + y_0\frac{\partial\phi}{\partial y} + z_0\frac{\partial\phi}{\partial z} + \frac{\partial\phi}{\partial t}\right)$$
$$+\rho\left(x_0\frac{\partial\phi}{\partial x} + y_0\frac{\partial\phi}{\partial y} + z_0\frac{\partial\phi}{\partial z} + \frac{\partial\phi}{\partial t}\right)^2 = 0 \quad\ldots\ldots\ldots\ldots(4).$$

The family of quadric surfaces of revolution round any line must include every family of concentric spheres with its common centre at a point on the line. Accordingly taking A_0 to be the origin, and $\phi(x, y, z)$ to be $x^2 + y^2 + z^2 + 1$, the family of spheres at any point (x_1, y_1, z_1) is included in the family

$$\lambda(x^2 + y^2 + z^2 + 1) + \mu(x_1x + y_1y + z_1z)^2 + 2\nu(x_1x + y_1y + z_1z) + \rho = 0,$$

that is, in the family

$$\lambda(x^2 + y^2 + z^2) + \mu(x_1x + y_1y + z_1z)^2 + 2\nu(x_1x + y_1y + z_1z) + \sigma = 0 \ldots(5).$$

For this is the family of quadrics of revolution round the line joining the origin to the point (x_1, y_1, z_1).

62. Consider any two infinitesimal projective transformations in the plane of xy. One transformation is defined by

$$\left.\begin{aligned}\frac{dx}{dt} &= a_{11}x + a_{12}y + a_{13} - x(a_1x + a_2y)\\ \frac{dy}{dt} &= a_{21}x + a_{22}y + a_{23} - y(a_1x + a_2y)\end{aligned}\right\}\ldots\ldots\ldots\ldots(1).$$

The other is defined by

$$\left.\begin{aligned}\frac{dx}{dt} &= b_{11}x + b_{12}y + b_{13} - x(b_1x + b_2y)\\ \frac{dy}{dt} &= b_{21}x + b_{22}y + b_{23} - y(b_1x + b_2y)\end{aligned}\right\}\ldots\ldots\ldots\ldots(2).$$

Now each of these transformations leaves a family of curves latent, the locus of points, which *either* are the points of contact of members of the respective families, *or* are points on a curve common to the two families, is given by

$$\frac{a_{11}x + a_{12}y + a_{13} - x(a_1x + a_2y)}{b_{11}x + b_{12}y + b_{13} - x(b_1x + b_2y)} = \frac{a_{21}x + a_{22}y + a_{23} - y(a_1x + a_2y)}{b_{21}x + b_{22}y + b_{23} - y(b_1x + b_2y)} \quad (3).$$

This locus is a cubic curve.

Now consider two rotations belonging to the congruence group under consideration. Let one be about the point (0, 0, 0), and the other about the point $(x_1, y_1, 0)$, and let the plane of xy be latent for them both. Then for the first rotation, the family of latent curves (cf. § 60, equation (1)) in the plane of xy is given by

$$\lambda(x^2+y^2)+\mu=0 \quad\text{.......................(4)};$$

and for the second rotation, the family of latent curves (cf. § 61, equation (5)) in the plane of xy is included in the family

$$\lambda_1(x^2+y^2)+\mu_1(x_1x+y_1y)^2+2\nu_1(x_1x+y_1y)+\sigma_1=0 \quad\text{......(5)}.$$

It is easy to prove that the locus of points where members of the family (4) touch members of the family (5) is the line, $x/x_1=y/y_1$.

Hence for the case of these two rotations the cubic curve of equation (3) above becomes a straight line and a common member of the two families (4) and (5). Thus these two families must possess a common member. Let it be

$$c_1(x^2+y^2)+1=0.$$

Then (cf. § 60, equation (1), and § 61, equation (5)) the sphere,

$$c_1(x^2+y^2+z^2)+1=0 \quad\text{.......................(6)},$$

belongs to the family of spheres centre (0, 0, 0), and also to the family of spheres centre $(x_1, y_1, 0)$.

Hence any two distinct families of concentric spheres with different centres possess one member in common.

63. Let (x_1, y_1, z_1) and (x_2, y_2, z_2) be any two points which are not collinear with the point (0, 0, 0). Let (cf. § 62)

$$c_1(x^2+y^2+z^2)+1=0 \quad\text{.......................(1)}$$

be the sphere common to the two families of spheres with centres at (0, 0, 0) and (x_1, y_1, z_1) respectively; and let

$$c_2(x^2+y^2+z^2)+1=0 \quad\text{.......................(2)}$$

be the sphere common to the two families of spheres with centres at (0, 0, 0) and (x_2, y_2, z_2) respectively. Then (cf. § 60, equation (3)) the family of spheres, centre (x_1, y_1, z_1), is given by

$$\lambda_1\{c_1(x^2+y^2+z^2)+1\}+\mu_1\{c_1(x_1x+y_1y+z_1z)+1\}^2=0 \quad\text{...(3)},$$

and the family of spheres, centre (x_2, y_2, z_2), is given by

$$\lambda_2\{c_2(x^2+y^2+z^2)+1\}+\mu_2\{c_2(x_2x+y_2y+z_2z)+1\}^2=0 \quad\text{...(4)}.$$

But (cf. § 62) it is possible to find a common member of the families (3) and (4). Then remembering that the two centres are not collinear with the origin, it is easy to prove that we must have

$$\mu_1=0, \quad \mu_2=0, \quad c_1=c_2 \quad\text{.....................(5)}.$$

Thus the three families of concentric spheres with centres at three non-collinear points have one member in common. Hence it is easy to prove that there is one sphere common to all families of concentric spheres. Let this sphere be called 'The Absolute.'

64. By a rotation round a suitable axis any point can be moved to any neighbouring position. For, if

$$c(x^2+y^2+z^2)+1=0 \quad\text{.......................(1)}$$

is the absolute, then (cf. § 63, equation (3))

$$c(x^2+y^2+z^2)+1-\{c_1(x_1x+y_1y+z_1z)+1\}^2=0 \quad\text{.........(2)}$$

is the equation of the sphere, centre (x_1, y_1, z_1), touching at the origin the plane

$$x_1x+y_1y+z_1z=0.$$

Hence if $x_1/l=y_1/m=z_1/n$, the sphere touches at the origin the plane

$$lx+my+nz=0.$$

Now let this be any plane through the origin and through the neighbouring position to which the origin is to be displaced. Then it follows that a rotation round a suitable axis through the point (x_1, y_1, z_1) can effect the required displacement of the origin.

Thus the effect of any infinitesimal congruent transformation can be produced by combining a rotation round some line not passing through the origin with a rotation round some line through the origin. Hence (cf. § 45, axiom (3)) the absolute is latent for any congruent transformation of the group.

65. Conversely the group of projective transformations, for which a given imaginary or convex quadric is latent, forms a congruence group. For take a tetrahedron, self-polar with respect to the given surface of the second degree, as the fundamental tetrahedron. Then the equation of the surface can be reduced to the form

$$c(x^2+y^2+z^2)+1=0 \quad\text{.......................(1),}$$

and, when $c=0$, the surface degenerates into the infinite plane.

The most general form of infinitesimal projective transformation is

$$\left.\begin{aligned}\frac{dx}{dt}&=u+a_{11}x+a_{12}y+a_{13}z-x(a_1x+a_2y+a_3z)\\ \frac{dy}{dt}&=v+a_{21}x+a_{22}y+a_{23}z-y(a_1x+a_2y+a_3z)\\ \frac{dz}{dt}&=w+a_{31}x+a_{32}y+a_{33}z-z(a_1x+a_2y+a_3z)\end{aligned}\right\}\quad\text{.........(2).}$$

This is to satisfy

$$x\frac{dx}{dt}+y\frac{dy}{dt}+z\frac{dz}{dt}=0 \quad\text{.....................(3)},$$

when (1) is satisfied.

Hence (3) becomes, after simplifying by (1),

$$cx(u+a_{11}x+a_{12}y+a_{13}z)+cy(v+a_{21}x+a_{22}y+a_{23}z)$$
$$+cz(w+a_{31}x+a_{32}y+a_{33}z)+(a_1x+a_2y+a_3z)=0 \quad\text{......(4)}.$$

Then (4) must either be identical with (1), or must be an identity. But it cannot be identical with (1). Hence it is an identity.

Thus $\quad cu+a_1=0, \quad cv+a_2=0, \quad cw+a_3=0,$

$$ca_{11}=0, \quad ca_{22}=0, \quad ca_{33}=0,$$

$$c(a_{12}+a_{21})=0, \quad c(a_{13}+a_{31})=0, \quad c(a_{23}+a_{32})=0.$$

Thus the general form of transformation is

$$\left.\begin{aligned}\frac{dx}{dt}&=u-\omega_3 y+\omega_2 z+cx(ux+vy+wz)\\ \frac{dy}{dt}&=v-\omega_1 z+\omega_3 x+cy(ux+vy+wz)\\ \frac{dz}{dt}&=w-\omega_2 x+\omega_1 y+cz(ux+vy+wz)\end{aligned}\right\} \quad\text{............(5)}.$$

But when the origin is fixed, these equations reduce to the equations (6) of § 58 for the general infinitesimal rotation round the origin of the corresponding congruence group. Also it is easy to see that the above equations give one and only one infinitesimal transformation which transports the origin to a given neighbouring point (udt, vdt, wdt), and at the same time transforms a given line l through the origin, and a given plane π through l, into a neighbouring line and plane respectively through the new position of the origin and the new position of l. Thus by § 64 and by axiom (3) of § 45 all the transformations of the form (5) belong to the associated congruence group.

Hence these equations give the general form of an infinitesimal congruence transformation, referred to a normal reference tetrahedron. The equation of the absolute is then

$$c(x^2+y^2+z^2)+1=0 \quad\text{.......................(6)}.$$

It follows from equations (5) by applying the 'Second Fundamental Theorem' (cf. § 36) that a congruence group is a six-limbed finite continuous group.

66. The congruence groups are divisible into three types.

In Type I, c is positive. Then the absolute (cf. § 65, equation (6)) is an imaginary quadric. The congruence axioms hold for the transformation of all points of the projective space by any members of such a congruence group. Such a congruence group is called Elliptic.

In Type II, c is negative. Then the absolute is a real convex quadric. The congruence axioms only hold for all points within the space enclosed by the absolute for transformations by any members of the corresponding congruence group. Such a congruence group is called Hyperbolic.

In Type III, the numerical value of c has diminished indefinitely. Groups of this type require further investigation. They are called Parabolic.

67. In the Parabolic case, when c diminishes indefinitely, the point equation of the absolute

$$c(x^2 + y^2 + z^2) + 1 = 0,$$

reduces to that of the infinite plane. Hence for every parabolic group a plane is latent.

Again in equation (3) of § 63 by putting $\lambda c = a$, $\mu = b - \lambda$, we find that the equation of any sphere, centre (x_1, y_1, z_1), can be written

$$a(x^2 + y^2 + z^2 - 2x_1 x - 2y_1 y - 2z_1 z) + b + c\{(bc - a)(x_1 x + y_1 y + z_1 z)^2 + 2b(x_1 x + y_1 y + z_1 z)\} = 0.$$

Hence when c diminishes indefinitely, and the coefficient of no term is infinite, the general equation for spheres, centre (x_1, y_1, z_1), becomes

$$a(x^2 + y^2 + z^2 - 2x_1 x - 2y_1 y - 2z_1 z) + b = 0 \quad \text{..........(1)}.$$

Hence every sphere cuts the infinite plane, which is latent for this special choice of coordinates, in the imaginary conic where

$$x^2 + y^2 + z^2 = 0,$$

cuts it. Thus this imaginary conic in the infinite plane is also latent.

Accordingly in the parabolic form the absolute is represented by the latent infinite plane, and by the imaginary latent conic in the infinite plane. A set of concurrent rectangular axes are a set of concurrent lines intersecting the infinite plane at the angular points of a triangle self-conjugate with regard to the absolute conic.

The general form of the infinitesimal transformation (cf. § 65, equations (5)), referred to a normal reference tetrahedron, reduces to

$$\left.\begin{aligned}\frac{dx}{dt}&=u-\omega_3 y+\omega_2 z\\ \frac{dy}{dt}&=v-\omega_1 z+\omega_3 x\\ \frac{dz}{dt}&=w-\omega_2 x+\omega_1 y\end{aligned}\right\}\quad\ldots\ldots\ldots\ldots\ldots\ldots(2).$$

CHAPTER VIII.

METRICAL GEOMETRY.

68. The theory of distance follows immediately from that of congruence by noting two facts. In the first place let the anharmonic ratio* of the range $(PQRS)$ be denoted by $\{PQRS\}$; then if

$$A_1,\ A_2,\ P_1,\ P_2,\ P_3$$

are collinear points, we have

$$\{A_1P_1A_2P_2\} \times \{A_1P_2A_2P_3\} = \{A_1P_1A_2P_3\},$$

or, in another form,

$$\log\{A_1P_1A_2P_2\} + \log\{A_1P_2A_2P_3\} = \log\{A_1P_1A_2P_3\} \ \ldots\ldots(1).$$

In the second place, let A_1 and A_2 be the two real or imaginary points in which the line containing the points P_1, P_2, P_3 meets the real or imaginary absolute of some definite congruence group. Then for any transformation of that group (α) the anharmonic ratios are unaltered because the transformation is projective, and (β) the points A_1 and A_2 are transformed into the points in which the transformed position of the line $P_1P_2P_3$ cuts the absolute.

Thus if some multiple† of $\log\{A_1P_1A_2P_2\}$ be defined as the distance between the points P_1 and P_2, where A_1 and A_2 are the points where the line P_1P_2 cuts the absolute, then equation (1) secures the characteristic addition property of distance in respect to collinear points, and the second consideration secures the characteristic invariability of distances in a congruence transformation.

* Cf. Proj. Geom. § 38.

† This definition is due to Cayley, *Sixth Memoir on Quantics*, Phil. Trans. 1859 and Coll. Papers, vol. II., and to Klein, *Ueber die sogenannte nicht-euklidische Geometrie*, Math. Ann. vol. IV. 1871.

69. Now let P_1 be the point (x_1, y_1, z_1), and P_2 the point (x_2, y_2, z_2). Then the coordinates of any point on the line P_1P_2 take the form

$$\frac{\lambda x_1 + \mu x_2}{\lambda + \mu}, \quad \frac{\lambda y_1 + \mu y_2}{\lambda + \mu}, \quad \frac{\lambda z_1 + \mu z_2}{\lambda + \mu}.$$

Thus the points A_1 and A_2, where the line P_1P_2 cuts the absolute,

$$c(x^2 + y^2 + z^2) + 1 = 0,$$

are given by the roots λ_1/μ_1 and λ_2/μ_2 of the quadratic equation

$$\lambda^2\{c(x_1^2 + y_1^2 + z_1^2) + 1\} + 2\lambda\mu\{c(x_1x_2 + y_1y_2 + z_1z_2) + 1\} + \mu^2\{c(x_2^2 + y_2^2 + z_2^2) + 1\} = 0 \quad \ldots\ldots\ldots\ldots\ldots(1).$$

For the elliptic case, when c is positive, put

$$\cos\theta = \frac{c(x_1x_2 + y_1y_2 + z_1z_2) + 1}{\{c(x_1^2 + y_1^2 + z_1^2) + 1\}^{\frac{1}{2}}\{c(x_2^2 + y_2^2 + z_2^2) + 1\}^{\frac{1}{2}}}.$$

Then* $$\{A_1P_1A_2P_2\} = \mu_1\lambda_2/\lambda_1\mu_2 = e^{2\iota\theta}.$$

Thus the distance P_1P_2, written dist (P_1P_2), can be defined by

$$\text{dist}(P_1P_2) = \frac{\gamma}{2\iota}\log\{A_1P_1A_2P_2\} \quad \ldots\ldots\ldots\ldots\ldots(2).$$

Hence

$$\cos\frac{\text{dist}(P_1P_2)}{\gamma} = \frac{c(x_1x_2 + y_1y_2 + z_1z_2) + 1}{\{c(x_1^2 + y_1^2 + z_1^2) + 1\}^{\frac{1}{2}}\{c(x_2^2 + y_2^2 + z_2^2) + 1\}^{\frac{1}{2}}} \ldots(3).$$

It is evident that there will be two distances, associated with the two segments into which the point-pair P_1 and P_2 divides the line P_1P_2. If one distance, say the smaller, is called dist (P_1P_2), the other will be $\pi\gamma -$ dist (P_1P_2). Thus the whole length of a straight line is $\pi\gamma$. This system of metrical geometry embraces the whole of Projective Space†.

70. For the hyperbolic case, when c is negative, put

$$\cosh\theta = \frac{1 + c(x_1x_2 + y_1y_2 + z_1z_2)}{\{1 + c(x_1^2 + y_1^2 + z_1^2)\}^{\frac{1}{2}}\{1 + c(x_2^2 + y_2^2 + z_2^2)\}^{\frac{1}{2}}}.$$

* Cf. Proj. Geom. § 38.

† The possibility of a Metrical Geometry with closed lines of finite length was first suggested by Riemann, cf. *loc. cit.* For a full account and amplification of Riemann's treatment of distance, cf. *Forms of Non-Euclidean Space*, by F. S. Woods, printed in *The Boston Colloquium*, New York, 1905.

Then if P_1 and P_2 are both within the region enclosed by the absolute, θ is necessarily real.

Hence (cf. § 69, equation (1))

$$\{A_1P_1A_2P_2\} = \mu_1\lambda_2/\lambda_1\mu_2 = e^{2\theta}.$$

Thus the distance P_1P_2, written dist (P_1P_2), can be defined by

$$\text{dist}\,(P_1P_2) = \tfrac{1}{2}\gamma \log \{A_1P_1A_2P_2\} \quad\text{.................(1).}$$

Therefore

$$\cosh \frac{\text{dist}\,(P_1P_2)}{\gamma} = \frac{1 + c\,(x_1x_2 + y_1y_2 + z_1z_2)}{\{1 + c\,(x_1^2 + y_1^2 + z_1^2)\}^{\frac{1}{2}}\,\{1 + c\,(x_2^2 + y_2^2 + z_2^2)\}^{\frac{1}{2}}} \quad\text{...(2).}$$

There will only be one distance between P_1 and P_2. This must be associated with the sole segment of the line P_1P_2 which lies wholly within the region enclosed by the absolute. This system of metrical geometry only embraces those points which lie within the region enclosed by the absolute*. Any point in the region to which the metrical geometry applies is at an infinite distance from every point on the absolute.

71. The parabolic formula for the distance, arising when c is indefinitely diminished, can be derived as a limit from either of the other two cases. Put $\gamma^2 c = \pm 1$, according as c is positive or negative, so that γ increases as c diminishes numerically. Then expanding both sides of equation (3) of § 69, or of equation (2) of § 70, and proceeding to the limit, we find

$$\{\text{dist}\,(P_1P_2)\}^2 = (x_1 - x_2)^2 + (y_1 - y_2)^2 + (z_1 - z_2)^2 \quad\text{......(1).}$$

The parabolic system of metrical geometry embraces all projective space with the exception of points on the latent plane, which is the infinite plane in our system of coordinates. This is the ordinary Euclidean Geometry.

72. Exactly the same procedure can be applied for the measurement of the angle between planes. Let p_1 and p_2 be any two planes, and let t_1 and t_2 be the two real or imaginary planes through the intersection of p_1 and p_2 and tangential to the absolute. When the

* Metrical Geometry of this Hyperbolic Type was first discovered by Lobatschefskij in 1826, and independently by J. Bolyai in 1832. This discovery is the origin of the modern period of thought in respect to the foundations of Geometry.

congruence group is elliptic, or when the congruence group is hyperbolic, and the line of intersection of p_1 and p_2 passes through the region enclosed by the absolute, then t_1 and t_2 are necessarily not real. Then the angle between the planes is defined to be $\frac{1}{2\iota}\log\{t_1 p_1 t_2 p_2\}$.

Thus if the two planes are given by

$$l_1x + m_1y + n_1z + p_1 = 0, \quad l_2x + m_2y + n_2z + p_2 = 0,$$

and θ is the angle between them, we have

$$\cos\theta = \frac{l_1l_2 + m_1m_2 + n_1n_2 + cp_1p_2}{\{l_1^2 + m_1^2 + n_1^2 + cp_1^2\}^{\frac{1}{2}}\{l_2^2 + m_2^2 + n_2^2 + cp_2^2\}^{\frac{1}{2}}} \quad \ldots\ldots(1).$$

As before, there are two angles θ and $\pi - \theta$; but it can be proved that the whole angle round a line is 2π, owing to the existence of diametrically opposite regions in the neighbourhood of the line.

In the parabolic case, when c is indefinitely diminished, the angle between the planes is given by

$$\cos\theta = \frac{l_1l_2 + m_1m_2 + n_1n_2}{(l_1^2 + m_1^2 + n_1^2)^{\frac{1}{2}}(l_2^2 + m_2^2 + n_2^2)^{\frac{1}{2}}} \quad \ldots\ldots\ldots\ldots(2).$$

73. The same procedure can also be applied for the measurement of the angle between two concurrent lines. Let l_1 and l_2 be any two concurrent lines in a plane p. Let t_1 and t_2 be the real or imaginary tangents from the point $(l_1 . l_2)$ to the conic which is the section of the absolute by the plane p. When the congruence group is elliptic, or when the congruence group is hyperbolic and the point $(l_1 . l_2)$ lies within the region enclosed by the absolute, then t_1 and t_2 are necessarily imaginary. Then the angle between the lines is defined to be $\frac{1}{2\iota}\log\{t_1 l_1 t_2 l_2\}$. Thus there are two angles θ and $\pi - \theta$ between two intersecting lines, and the whole angle round a point is 2π.

In the degenerate parabolic case the section of the absolute by the plane p becomes two conjugate imaginary points in the plane at infinity. These are known as the circular points at infinity. Then t_1 and t_2 are the imaginary lines from the point $(l_1 . l_2)$ to these points respectively*.

* This projective view of Euclidean Metrical Geometry was elaborated by Laguerre in 1853, previously to the rise of the general theory which is explained here.

74. Thus Metrical Geometry is in fact the investigation of the properties of a particular congruence group. Any set of axioms of congruence (cf. §§ 43 to 45) form the definition of what we mean by a congruence group. The investigations which are summed up in equations (5) of § 65 and in equations (2) of § 67 form the proof of the existence of congruence groups in a projective space for which the axioms of order and of Dedekind continuity hold. It is proved that to any convex quadric and to any imaginary quadric with a real equation exactly one congruence group corresponds. Also there is one congruence group corresponding to each imaginary conic lying in a real plane and defined by a real equation.

If the absolute is a real quadric, the metrical geometry applies only to the region within it. If the absolute is an imaginary quadric, the metrical geometry applies to all the projective space. If the absolute is an imaginary conic in a real plane, the metrical geometry applies to the whole of the projective space with the exception of the real plane.

75. It follows that in relation to Projective Geometry no additional geometrical axiom is required in order to establish metrical properties. But the case is otherwise in respect to Descriptive Geometry. The transformations of a congruence group in Descriptive Geometry are to be one-one transformations of descriptive points into descriptive points, and all the other axioms of congruence can be enunciated without change of form. Thus when the associated projective space is formed, associated congruence groups in the projective space must exist, which however satisfy the further conditions (1) that proper projective points are to be transformed into proper projective points and (2) that the congruence conditions are to hold throughout the whole region of the proper projective points.

It follows therefore that the convex boundary surface of the proper projective points (cf. § 30) must be a quadric surface, or in the degenerate case a real plane. Unless this is the case no congruence group can exist in the original descriptive space.

Thus the Euclidean axiom (cf. § 10) is sufficient to secure the existence of parabolic congruence groups having as their latent plane the single plane of improper projective points (the points at infinity). Also with this axiom no other types of congruence groups can exist. But it is to be noticed that alternative congruence groups exist, namely one for each imaginary conic lying in the plane at infinity.

In order to secure the existence of hyperbolic congruence groups an axiom is required which secures that the boundary of the proper projective points is a quadric. Then it is to be noticed that one and only one congruence group exists in the descriptive space, namely that one which corresponds to this definite quadric. Perhaps the most direct form of the axiom is to assert that a hyperbolic congruence group exists.

Cambridge Tracts in Mathematics and Mathematical Physics

No. 1. VOLUME AND SURFACE INTEGRALS USED IN PHYSICS, by J. G. LEATHEM, M.A. *2s. 6d. net.*

No. 2. THE INTEGRATION OF FUNCTIONS OF A SINGLE VARIABLE, by G. H. HARDY, M.A. *2s. 6d. net.*

No. 3. QUADRATIC FORMS AND THEIR CLASSIFICATION BY MEANS OF INVARIANT FACTORS, by T. J. I'A. BROMWICH, M.A., F.R.S. *3s. 6d. net.*

No. 4. THE AXIOMS OF PROJECTIVE GEOMETRY, by A. N. WHITEHEAD, Sc.D., F.R.S. *2s. 6d. net.*

No. 5. THE AXIOMS OF DESCRIPTIVE GEOMETRY, by A. N. WHITEHEAD, Sc.D., F.R.S. *2s. 6d. net.*

In preparation.

THE DEFINITE INTEGRAL, ITS MEANING AND FUNDAMENTAL PROPERTIES, by E. W. HOBSON, Sc.D., F.R.S.

SINGULAR POINTS AND ASYMPTOTES OF PLANE CURVES, by Miss C. A. SCOTT.

THE EIKONAL AND ITS APPLICATION TO OPTICAL INSTRUMENTS, by E. T. WHITTAKER, M.A., F.R.S.

THE FUNDAMENTAL THEOREMS OF THE DIFFERENTIAL CALCULUS, by W. H. YOUNG, Sc.D.

THE CIRCLE AT INFINITY, by J. H. GRACE, M.A.

Made in the USA
Monee, IL
04 November 2026

67383825R00049